T0304391

SOMETHING IN THE
WOODS LOVES YOU

SOMETHING
IN THE WOODS
LOVES YOU

JAROD K. ANDERSON

Timber Press
Portland, Oregon

Note to the reader: This book explores depression and suicide.

Timber Press
Workman Publishing
Hachette Book Group, Inc.
1290 Avenue of the Americas
New York, New York 10104
timberpress.com

Timber Press is an imprint of Workman Publishing, a division of Hachette Book
Group, Inc. The Timber Press name and logo are registered trademarks
of Hachette Book Group, Inc.

Third printing 2025
Printed in the United States on responsibly sourced paper

Text design by Adrianna Sutton & Leigh Thomas
Cover image copyright the Artist (Tuesday Riddell), reproduced with
grateful thanks to MESSUMS ORG. Photo: Steve Russell.
Jacket design by Leigh Thomas

The publisher is not responsible for websites (or their content)
that are not owned by the publisher.

The Hachette Speakers Bureau provides a wide range of authors
for speaking events. To find out more, go to hachettespeakersbureau.com
or email hachettespeakers@hbgusa.com.

ISBN 9-781-64326-229-1

A catalog record for this book is available from the Library of Congress.

CONTENTS

WINTER

IN OHIO, WINTER IS LANDSCAPE POETRY. A white page. An elm scribbled on a snow hill. Empty space making each syllable of life more vital.

Yet poetry, like most magic in art and nature, asks you to meet it halfway.

When I stepped away from my career, in the frozen dark of chronic depression, I was not interested in magic. I was interested in sleep. In escape. In seeking an end.

The bare trees. The gray sky. The stillness of absent life.

These things are the surface of winter.

But winter isn't an ending.

Winter is the deep breath before a song.

Chapter 1

BLUE HERON

THERE'S AN OLD STORY ABOUT GREAT BLUE HERONS. It says that while hunting the twilight shallows, herons can produce a strange, luminescent powder, pluck it from between their feathers with their spear-like beaks, and sprinkle it on the dark water to attract fish.

Picture it.

Iridescent bluegill rising up beneath the surface, drawn to sudden, phantom starlight beneath the shadowed canopy of a great bird's wings. The motes of cold fire sparkling in their unblinking eyes. Above, beyond the glimmer, a beak like a frozen thunderbolt paints a dull, gold slash in the air.

The fish are not curious in an intellectual way. It's a physical thing, their bodies called forward to witness the inexplicable. There, in the shallow winter waters, they are ready to believe in miracles.

This old story is a myth, but it's not hard to imagine why such a story gets passed on. It tells a figurative truth within a literal falsehood, a pathway to a kind of knowledge.

Yes, technically speaking, it's a lie.

Technically speaking, you can look at any human life as the sum of a complex collection of chemical reactions, in much the same way as you can look at any beautiful painting as a simple collection of pigments, which is to say, you can miss the point of anything.

Some herons migrate, but here in Ohio, I tend to see them all winter long, tall and solitary, moving with a deliberate slowness that complements the placid waters in which they wade. They look pensive and intense. Thin wanderers in rags divining the future by studying ripples on the leaden waters of January. They are a mix of shaggy and angular, a blade of yellow stone dressed in flowing robes stitched from overcast skies.

From beneath, a fish's perspective, they are the pale hue of heavy afternoon clouds. From above, the darker shade of flinty shallows. All of this, the slowness, the camouflage, the living statue in the stillness, it all blooms outward from that long, yellow beak like a sunset dagger poised in the air.

Herons are gray wanderers. They are shards of winter landscape. They are the sky from below, and, from above, they are the dark water regarding itself. No, they don't produce glowing dust, but if you don't think herons are magic, I suggest you need to broaden your definition of that word.

They can symbolize fear or gluttony or grace or patience.

It depends on who you ask.

I have difficulty interpreting the nature of my own life, a thing I feel intimately and continuously, so it's not surprising that we can't all agree on the nature of herons.

What, then, is a blue heron?

They live about fifteen years. They stand around four feet tall. They walk the shoreline, delivering an overt message to me about quiet contemplation and self-determination.

The heron is exactly what the heron is to you in the moment you choose to give it meaning. It will be that meaning until you decide it means something else. That's how meaning works. It's a subjective act of interpretation.

You might get the impression that I'm saying herons are meaningless, but that's not what I'm saying at all. When I see a heron and interpret its behavior as a reminder for me to slow down and think about what actually matters in my life, that is what that heron means. Meaning, like many crafts, happens in collaboration between maker and materials.

There's nothing nihilistic about this.

A human alone in the wilderness has neither fur nor fangs but does have the ability to build shelter. Our nature understands that we must build what we need. Community. Tools. Shelter. Meaning.

The heron allows me to build the meaning I need for the moment I need it. Making meaning in this way is like creating harmony with two voices. I sing my portion. The heron sings hers. The harmony is woven and meaning exists in the world.

The trouble starts when we forget about our participation in the creation of harmony, of meaning. When we remove our agency in meaning-making, we start to think in absolutes. We are our jobs. We are defined by the car we drive. Our objective worth is reflected in the way our parents or our peers treat us. We don't lend our voices to harmony. We buy our harmonies pre-sung in tidy plastic packages.

I believe our worth is fixed. We are innately worthy. Our meaning? Our identity? Those concepts require our intentional participation and they are mercifully flexible.

The heron only represents self-determination when I need her to. That doesn't diminish the heron's power. It simply highlights my own.

There are objective facts in the world. Of course there are. But our concept of self, our significance, our sense of whether or not we deserve to take up space in the universe or experience joy and contentment—these are not questions of fact, they are questions of meaning.

The heron sings her portion of the harmony. We sing ours.

Together, we make something new.

Something we dearly need.

❖

After graduate school, armed with a master's in literature, a loving partner, immense debt, brutal and unacknowledged mental illness, and no plan beyond escaping academia, I entered a grim half dozen years during which I don't recall ever seeing a heron.

The truth is, I must have seen them. They are dinosaurs when in flight and fairytale creatures when walking the river. Hard to miss. But, somehow, I never noted their presence.

During those years, I was not concerned with herons.

I entered a period in which the world had stopped being a source of wonder. The world was something to fend off. Nature was a ghost in the attic, a memory of childhood. It wouldn't pay my bills or quiet my persistent urge to end my life. A heron was a half-remembered song from a dead age, a soiled stuffed animal wedged in a storm drain. In those grim years, a wading bird didn't merit interpretation.

Eventually, I found my footing in marketing and nonprofit work. A series of increasingly solid jobs that saw me become better paid and increasingly hopeless. More respectable and less in touch with my own mind and body. I worked as a coordinator. Then a manager. Eventually holding the title of Director of External Relations for Ohio University's Zanesville campus.

I was doing the things I was supposed to do.

I was climbing the ladder.

After my master's, I'd been accepted to a PhD program, but by that time I had a clearer picture of what a life spent living on $1,200/month stipends for another four years (and for an unspecified number of years after that as I moved around the country fighting for increasingly elusive teaching positions and living on adjunct wages below the poverty line in a time of shrinking budgets and closing campuses) would look like. So, I turned down the PhD and gambled that wherever I worked could hardly be more exploitative than academia.

In a relatively short number of years, I was back in a university context, but on the administration side. I had a nice office. I was paid on par with tenure track professors. I was told I would even have the opportunity to teach if I liked, all without enduring years of starvation-wage adjunct positions in the common, current fashion of building a career on the teaching side of academia.

In many ways, I felt like I had outsmarted the system.

By the time I held the director title at Ohio University, I had an impressive resume and unchecked chronic depression that led me to contemplate suicide several times an hour.

In the end, desperate and shattered, but with the unyielding support of my wife, Leslie, I made the terrifying decision to leave work and try to rebuild my mental health. The herons had gone from my life, and I didn't even know to miss them.

I had no plan.

I had no hope.

I carried immense shame about being a burden to my partner.

I knew, intellectually, that I was profoundly fortunate to have the space and support to prioritize my health and recovery. I had a rare privilege. I knew it, but I didn't feel it. I felt hollow and miserable, more so as I understood that ending my career didn't end my pain. I felt as if I had escaped from a burning building only to find that the entire world was burning. There was no real escape to be found in life.

I just wanted to die.

Then, two months after leaving the job I had expected to be my career for the next thirty years, I took my first intentional walk through the woods. Nature may have felt like a relic of my distant past, but it was a relic I could visit in the waking world, a memory with a parking lot.

It was January, a Wednesday morning. Ohio was wearing its traditional gray on gray on gray. Gray skies. Gray roads speckled with salt residue. Ranks of gray trees, bare and silent.

Chronic major depression had been slowly killing me for years. I had disassociated away vast swaths of memory. I was adrift without a career or a goal or a purpose, but a landmark from childhood loomed large within my skull and I made for it. The woods. The quiet beneath the trees. The pure, intoxicating pull of small rivers and flat stones that dare you to look beneath.

So, I went to the trees and the water. I drove a little way south to a park called Shale Hollow. It isn't a big place. A patch of protected wilderness not far from a busy highway crowded with shopping centers. A shallow creek winds through the park and a few tall shale formations show off perfectly round concretions, spherical carbonate rocks bigger than beachballs exposed between the finely stacked layers of stone.

I remember climbing out of my warm car and smelling the citrus tang of leaf rot and muddy water. It was like seeing an old friend across a restaurant, feeling a draw to speak with them and a pang of fear that reconnecting after so many years was impossible.

What did I have to say to the wilderness? What did we still have in common?

Somewhere distant, I heard children laughing beneath the trees.

A crow cawed and was answered.

I shut my car door and walked for the nearest path.

13

I crossed a footbridge, dark planks marbled with green and grime, and entered the woods beneath the skeletal branches of winter-bare trees. To my right, a leaf-strewn path of hardpacked earth disappeared up a rocky slope. I followed it.

I felt odd.

Alone in the woods, I was different.

I wasn't an employee.

I wasn't a spouse or a son or a student.

I wasn't friend or teacher.

I wasn't patient or client.

I was just there.

I had no cultural or social role to perform.

Just me.

An animal.

A collection of senses.

No one was asking what I did for a living or waiting for me to make a witty observation.

No one was giving me a sales pitch or expecting one from me.

It was disorienting, both in terms of what was there and what was not.

I took careful steps, pressing my face close to trees, studying the otherworldly maps of lichens on the bark, strange continents in yellow, green, and brown. I wondered about the names of this cluster of pines or that patch of dry moss. I used to know these things. It was like walking into my old elementary school and feeling that vertiginous sense of everything being the same, yet utterly different. It was a bit nostalgia and a bit déjà vu.

"That tree with strips of bark hanging like sheepdog's fur. I knew its name once."

"That woodpecker, small and quick, dappled black and white. I knew its name once."

I walked, wondering about things I had forgotten, feeling my emotional pain hot on my neck like a second sun.

My depression often manifests as self-criticism, as "should" thoughts. I should be eating better. I should be contributing more to my family. I should be doing more to help, to earn, to create. Yet there, in those woods, there were no vital tasks I was meant to perform. No fleeting chance at prestige or material gain. I drove there for the express purpose of being there, and it was hard to argue that I was failing at that simple goal.

I was conscious of my overwhelming ignorance about what I was seeing, but it's not as if the trees are insulted when you don't know their scientific names. There was no test waiting for me in the parking lot. My vague internal fear that I was "doing the woods incorrectly" couldn't provide any supporting evidence.

My path brought me alongside a broad, shallow stream, its banks piled with fractal shards of dark shale that shone wetly in the morning light. I looked upstream, then downstream. There was no one. The distant sounds of traffic murmured somewhere beyond the trees, a conversation in the next room.

My shoes crunched on the flaking stone that covered the banks like spilled puzzle pieces. The wind stung my face. The water smelled like soil and decay. Every bird call was a voice I had once known, a reminder of a forgotten time.

Then I saw the heron.

A great blue heron standing still as a statue on the edge of a pool. She saw me. She wasn't fishing, just standing like a picture in a book, her yellow eye fixed on my face. I matched her stillness. She looked too big, too unlikely, too strange to be standing out in the open for anyone to see.

Yet, she was there.

And I remembered.

I knew this bird.

I knew her name. I knew the way she folded her neck as she flew, the way that beak shot toward the water like an arrow from a bow. I knew this bird and, in that moment, I knew something else.

It's all still here, I thought.

The heron. The creek. The trees.

It's not dead and gone.

It's not locked away in the past.

It's not a phantom of childhood or a metaphor for a bygone age.

It's still here.

A hundred thousand little rivers. A million secret sights beneath fallen logs and riverbank rocks. It was all still here.

I had gone away, not nature.

But, of course, that's not quite right either.

I hadn't gone anywhere.

I had just stopped noticing.

There are two paths to magic: Imagination and paying attention. Imagination is the fiction we love, the truths built of falsehoods, glowing dust on the water's surface. Paying attention is about intentional noticing, participating in making meaning to lend new weight to our world. An acorn. The geometry of a beehive. The complexity of whale song. The perfect slowness of a heron.

Real magic requires your intention, your choice to harmonize. Of course it does. The heron cannot cast starlight onto the dark shallows to entrance the bluegills. Not unless you do your part. You must choose to meet her halfway. And when you do, you may find that magic isn't a dismissal of what is real. It's a synthesis of it, the nectar of fact becoming the honey of meaning.

I stood there, watching the heron, watching as she decided I was not a threat, as she turned and paced upstream, a poem of ancient slowness. I stood and I silently cried, with the wind numbing my ears and my tears shining in the sunlight.

I was a grown man, alone in the woods, crying at a heron's back, and even if I couldn't articulate it at the time, it was the most hopeful I'd felt in years.

There was the bird, but not just a bird.

The encounter was a bridge to when nature was family.

It was my storybook wizards.

It was science as magic, nature as art.

It was priceless and essential, yet it had no use for my money or imagined prestige.

I could feel that the heron was alive in the same way I was alive.

I couldn't deny that the world that made the heron made me too, that we were of the same time and context, and somehow, in that moment, it didn't feel possible that I was made to be miserable and afraid, to be measured by bank statements or resumes, not when there were living poems, descended from dinosaurs, walking beneath the silent trees just as they had long before the first written language.

The heron told me that my better days were not beyond reach and that the world was more than pain, bitter news, and sleepless nights.

There are wonders here.

Things worth experiencing, worth knowing.

Magic hidden in plain sight.

Bats can hear shapes. Plants can eat light. Bees can dance maps. We can hold all these ideas at once and feel both heavy and weightless with the absurd beauty of it all. These are some facts that are easy to overlook.

These are facts that saved my life.

It would be reductive and dishonest to say that my road to mental health was as simple as inviting nature back into my life. It wasn't. Yet, reconnecting with the natural world was a key step along the way. Often, to heal, we need a reason to seek healing. Depression tells us there are no such reasons, that healing is hard and not worth the effort. Nature told me something different.

Chapter 2

GRAY SQUIRREL

THE LESSON OF THE HERON LINGERED.

February approached.

I doubted my budding interest in nature, the old critical voice in my head calling it "childish" and "frivolous." It didn't matter. I had learned something and that something had sunk its roots deep. Perhaps I had been on the wrong path in life, but if so, I had discovered that old paths were not so impossible to revisit. Afterall, I had revisited them. Literally.

I wanted to invite nature back into my life, but I wasn't sure how. I wanted to reconnect with something real and fundamental. So I turned to the default tactic of my culture when approaching any new endeavor. I surrendered to my training. I bought things.

I found a small, boutique birdfeeder store and spent money I didn't have on heavy-duty bird-feeding equipment. I was serious about connecting with nature. I had the receipts to prove it.

In the fancy birdfeeder store, I learned about feeding songbirds and excluding everything else. I learned about invasive species like the European starling, how they were brought to America, along with house sparrows, by a misguided group in nineteenth-century New York who wanted the United States to host all the birds mentioned in Shakespeare's plays. I also learned how to thwart them at the feeder, with safflower seeds that are too tough for their slender beaks to open.

The bird store warned me about pest animals like squirrels. Especially squirrels. From pole baffles to cayenne-pepper-laced seed mixes, there is a whole range of products designed to defend against squirrels. The marketplace was telling me a clear story in the language of inventory, a story about which animals were desirable backyard guests and which were not.

I bought three feeders, a pole, a baffle, and plenty of starling- and squirrel-resistant food.

I erected my feeding station despite the frozen ground and waited for my purchases to do their work. Step one: spend money. Step two: let the nature connection begin.

I'm being flippant, but the truth is, it was a start.

In the depths of depression, most any action in pursuit of any goal is a powerful gesture of defiance against that inner voice telling you to sit still with your misery.

There I was in my quiet house, wounded and uncertain, watching cardinals and chickadees come and go. My new identity. An unemployed bird benefactor. A cunning opponent of starlings and squirrels. A guy who saw a heron and bought pricey birdfeeders.

Beyond questions of identity, a key unknown paced like a caged bear on the periphery of my thoughts.

Was it enough?

Was I finding reasons to keep stringing days together?

Was this what I turned away from my career to do?

Was I healing?

We understand our world and ourselves through stories.

Conflict.

Character growth.

Protagonists and antagonists.

Few are the stories we tell about patient acceptance of our limitations or careful cultivation of contentment.

My long years of depression have worn certain regions of my memory as smooth and featureless as river stones, no doubt robbing me of many stories I might mourn if I could sense the shape of their absence. Many of the years leading up to those uncertain moments watching the birds were touched by those scouring waters. From the time I spent as a student and employee at Ohio University, only a handful of stories stand out, dark silhouettes like tall pines in a rolling tide of fog.

One of them is about a squirrel.

❖

In my memories of Athens, Ohio, the brick streets are always wet with rain. It's an unseasonably warm winter day that smells like overripe autumn. The sycamores gleam like bleached bones jutting from the hilltops, bright and distinct among the sober university halls, ochre and tan, bronze and weathered concrete.

On one such day, the one endless day of my Athens memories, I was walking to my English Department office in Ellis Hall when I noticed a line of halted traffic. I couldn't see why they were stopped until I reached the front car. I made eye contact with the driver. She looked to the road and I followed her gaze.

There on the wet bricks was a gray squirrel. Tail down. Limbs tucked inward. Seemingly awake and aware, but stone still.

The driver gave me a micro-shrug and continued to wait for the squirrel to do whatever it would do. Somewhere down the line of cars, a horn blasted. An angry voice rose and fell.

On impulse, I did something foolish.

I walked up to the squirrel.

She didn't move.

I reached down and laid a hand on the squirrel's back. She was warm and her heartbeat drummed against my fingertips.

She still didn't move.

I scooped her up with two hands and carried her out of the street.

In the years since that day, I have been acquainted with wildlife rehabbers who have puckered scars from squirrel bites, but I had no such experience then.

Behind me, I heard traffic resume, tires muttering on slick brick.

The squirrel was a soft furnace, vibrating like a tiny engine.

I considered what I knew about tending to injured wildlife, which was nothing at all.

The road, University Terrace, was cut into the hilltop, hemmed in by sidewalks and a low stone wall beyond which was a vast lawn. I don't recall anyone commenting on my actions, but Athens is an odd place, a small, Appalachian college town where a man carrying a squirrel might not rate a second look from passersby. I placed the squirrel carefully in the grass, expecting to spend the next few minutes as an uncomfortable witness to the small creature's final moments, standing sentinel as whatever injury had stilled her body finished taking hold.

I withdrew my hands and stood.

The moment I did, the squirrel rose and raced off toward the nearest tree. A sprint and a leap, her plane of motion shifting from horizontal to vertical as she skittered up the trunk, then vanished behind the thick boughs.

I wondered at her experience of events, imagining her perspective.

The fear in the road. The inscrutable cars. The touch of human skin. The return to grass and the frantic moment of escape. A rush, an arc through the air, the flat, winter-sparse horizon transitioning to a branching highway stretching up toward limitless sky.

I don't know if the squirrel had fallen and was momentarily stunned. I don't know if she visited the road to lick salt and lost consciousness to some unguessable malady. All I know is that the change in position, brick to grass, a return to a familiar context, unlocked whatever mechanism had held the creature motionless.

We understand the world through stories.

The stories we hear of others, role models and cautionary tales. The stories we tell about ourselves to ourselves, the constantly unfolding saga of identity. But what of the squirrels? They clearly function just fine in a world without human narrative. Is it possible for us to step outside our own stories and imagine their worldview?

I will never comprehend a tree as a footpath like a squirrel can. I can't understand the sky the way a vulture riding a thermal does. I can't know what a pond is the way a musk turtle knows. But I will sense the presence of these hidden perspectives and feel grateful that my human ways of knowing the world do not hold a monopoly on reality. There's comfort in my certainty that the scope of natural perception is not tethered to my own limitations.

Perhaps it's worth imagining ourselves from the perspective of squirrels.

How brave we are to walk boldly across the ground in full view of hawks and foxes. How impossibly giant. How remarkably long-lived. How magnificent it must feel to be too heavy to be carried off by grasping talons and beating wings.

We can all imagine the biting words of make-believe critics. So, we can also imagine the accolades of admiring squirrels. We are, after all, storytellers. Make sure that some of the stories you tell yourself are kind.

I don't know why my memory of carrying the squirrel away from the road lingers on so vividly when so much from those years is faded. I don't know why, but I choose to see it as meaningful. A signpost. The restorative power of setting the squirrel in a new place and context. An act of reckless kindness and warm fur against my palms.

Perhaps gray squirrels are pests if you just spent $30 on birdseed in the hopes of an uninterrupted month of songbird visits, but in the broader ecological sense they are absolutely vital.

Stories.

One story of squirrels casts them as birdseed thieves, as powerline chewers, as interrupters of dog walks and invaders of attics.

There are others to tell.

❖

I live in a small, white house between a wooded park and a cemetery. The cemetery is a place of gentle quiet. At night, I can hear the whistle of distant trains. It's a lovely place.

The cemetery is quite old, for America. Revolutionary war veterans are buried there along with veterans of every American war since. I don't prefer to measure time in wars, but gravestones often do. The land doubles as an arboretum with mature trees, their species labeled with little silver tags. There are many oaks, hickories, and walnuts in the area. It's a place well suited for eastern gray squirrels to thrive.

In my part of the world, squirrels are the sort of commonplace wildlife that seems to disappear through familiarity. Walkers along a hiking trail might stop to watch a red-tailed hawk surveying a field or deer flicking their snowy tails as they flee into the brush, but few break stride for a squirrel. As with most invisible wonders, the problem is that we don't take the time to notice, to interrogate our typical viewpoints about squirrels. If we don't willfully "opt out," we default to the standard stories of our culture.

Yes, they will empty your birdfeeder. They are also the most effective and essential forest regenerators in North America, compulsively burying nuts, many of which they never retrieve. Some assume this is an accident, but I doubt "accident" is the correct word for an ancient process that expands forests and enriches species beyond count. It's amazing what humans will dismiss as happenstance in the absence of an anthropocentric story of clear intention.

Some gray squirrels have been observed pretending to bury a nut if they think they are being observed, a feint to fool would-be robbers. They are one of a very few mammals that can climb down a tree headfirst. They can live two decades, leap ten times their body length, transform the landscape and, ultimately, the climate. Yet, few people will pause to give one a second look.

The relationship between squirrels and trees is harmonious in a way that is lovely to consider. Instinctual gardeners living in their generational gardens. In the dynamic of squirrels and trees, it's hard to say which is the shepherd and which is the flock. There's a wonderful lesson in that, disparate creatures thoughtlessly caring for one another, as natural as breathing.

I've never been a religious person, but two places feel consistently holy to me. Forests and libraries. Given this, I suppose squirrels should count as a kind of cleric. It would be more fitting than their current place in our narratives as haphazard pests and suburban invaders. Maybe the bringers of forests are unwelcome in our enclaves of manicured lawns, but I'd suggest it's time we rethink what they are offering to us.

Our stories shape our worldview in powerful and invisible ways. Many of our most powerful stories gain their potence from simplicity. Good and evil. Right and wrong. Easy, bite-size narratives that spare us troublesome gray areas.

Squirrels are pests.

Shakespearean birds are quality birds.

The mentally ill person who struggles to leave the house can no longer be a protagonist in the central story of our world.

We all consume so many purposefully crafted stories that it's easy to forget life doesn't follow conventional narrative structure. We can't wait for our climax. We don't have character arcs. We live and then we don't. There is no final culmination in success or failure. We are not curated collections of achievements or mishaps.

It is not inherently wrong that we understand the world through stories. It's who we are. It is wrong not to treat our stories as living documents, as ongoing conversations. We know that our stories tend toward dichotomous patterns of black and white, so it's those structures that we need to scrutinize most actively.

This isn't easy advice to take. I still struggle with it. Stories are like gravity. It's hard to examine something you've felt your entire life.

❖

Sitting at the dining room table in my dark and silent house, watching the birds crack sunflower hulls, I couldn't summon the language for what I needed to do. I only felt the wrongness of the situation. I was desperately ill, and I knew on a deep level that my illness was not acceptable. I had stepped over a line in the narrative, and I was on the wrong side of the story. An antagonist. A nuisance animal. A verminous squirrel. And I knew, without a doubt, that my culture had many baffles to thwart me.

In the society in which I was raised, you must be incredibly and obviously ill before it is socially permissible to make the pursuit of healing your primary vocation. Permissible perhaps, though not often feasible. Among those who reach this arbitrary benchmark of illness, it remains a rare privilege to be able to walk away from other obligations, even when the stakes are life and death. Polite society in modern America does not abide public illness cheerfully, nor does it readily forgive its limitations.

These practical and cultural challenges are further complicated if the illness is too invisible or too conspicuous. Further still if it makes others uncomfortable or if it requires specialized equipment or rare expertise. The complications increase exponentially for those who do not have the emotional and financial support systems that I enjoy, those, for example, who navigate the labyrinthian regulations of federal disability programs, where funding can be stripped away for such missteps as finding someone you wish to marry or saving too much money. Our legal policies surrounding disability funding carry a clear message. If you need your civilization's help to stay afloat while disabled, you must be careful to live in the abject poverty society feels you deserve or the help you need will be withheld. Such is our cultural love of billable productivity and our general disdain for everything else. It's a concept that many of us internalize without a second thought. Our worth is our productivity.

I am no expert on disability nor can I speak for all sufferers of mental illness. I can't know the struggles of other chronic illness sufferers any more than I can perceive a branching oak the way a squirrel does, but I plant my empathy in the soil of my own pain and hope the love that grows there is enough to counterbalance my ignorance.

These ideas of productivity and worth drift like embers on the wind, spreading wildfire in young minds. We seem to understand innately that a five-year-old has fundamental worth that has nothing to do with their earning potential, but our ability to apply this knowledge to ourselves often wilts and dies. I, certainly, believed that I had become fundamentally broken when I no longer had a steady income stream.

The story archetype for a life well-lived in modern America is the story of steady wealth accumulation and a clear, upward career trajectory. I followed this narrative. It was my North Star. Even when I pursued a life of books and ideas, the unspoken goal still involved visible success and social legitimacy. These guiding stories led me to a life where mental illness was a coiling monster hastily nailed beneath the floorboards and suicide seemed preferable to deviating from my expected storyline.

There was a subtle twist to friends' and coworkers' faces when I'd hint at struggles with mental illness. A twist that said, "That's not one of our agreed upon topics." It said, "You're breaking the social contract. We don't look beneath those floorboards. That scratching is just a squirrel. It's always just a squirrel. Filthy intruders. If you just work hard enough, there won't be time to hear things like scratching beneath the floors."

These were not bad or cruel people.

They, like me, had adopted our commonplace stories as a kind of shorthand for moral understanding.

Especially when encountering issues we've never faced, it's easy to let prevalent narratives stand in for our own ideas and values. We live with so many pressures, making room for imaginative empathy can be a daunting task.

Mental illness should be hidden.

Worth comes from productivity.

Squirrels are purposeless pests.

These are popular, wrong ideas that can seem harmless if you happen to be in a position to ignore the harm they do.

A chickadee landed on my feeder, black, white, and gray. She plucked up a single sunflower seed and flew away to eat in private. A squirrel leapt lightly from the cemetery fence onto a blue spruce at the edge of my yard. I saw her study my feeder. She looked at the stovepipe baffle that would prevent her from climbing the pole. A pause. A flick of a tail. Then, she retreated back to the cemetery maples.

I felt the soft weight in my hands again as I walked from the road in Athens. I saw the moment when the squirrel sprang to life, when she raced up a tree which may well have been planted by her own ancestor. Squirrels are not pests simply because we create unwelcoming contexts for them. Illness is not a character flaw simply because we engineer needless burdens for the ill. Stories are meant to change in the telling.

The next day, I made another purchase.

I bought a squirrel feeder.

It was a small token of appreciation and apology.

It was a start.

Chapter 3

SUGAR MAPLE

THE LEAF OF A SUGAR MAPLE has always reminded me of a hand, five fingers spread in a gesture of greeting. In the deep of winter, there were no green hands waving hello from the bare maple that stood on the cemetery side of my back fence and acted as doorstep to the squirrels visiting my feeder. Yet, there were leaves visible high in the tree. A clump of them thirty feet up in the branches, the color of strong tea, a gray squirrel nest known as a drey, a shelter from winter winds, built of woven twigs and packed leaves. Gray squirrels don't hibernate, but the tree lends its hands to shelter its warm-blooded partners even as it dreams away the winter.

In those early days of my healing, I was spending most of my time sheltering in my own drey, waiting for some instinct to awaken in me and set my feet on the path to recovery. I was doing what I could not do while working fulltime, what I fantasized about doing while shut up in offices and meetings. I was doing nothing, and every moment of it was increasing the weight of the burdensome knowledge that doing nothing, that being free from obligation, was not the cure I'd imagined it would be.

Simply not going to work didn't generate a new perspective, nor did it grant me an effortless surplus of self-love and tranquility. It was terrifying to acknowledge, but my work life was *a* problem, not *the* problem.

There is meaning in the graceful stride of a heron.

There is worth in the daily lives of squirrels.

These ideas felt natural and inviting when I encountered them in the wild. The tension arrived when I tried to apply the same level of love, of interest and empathy, to my own life. I now had the time to be kind to myself, but I didn't have the capacity.

I understood how squirrels could be worthy representatives of nature. Their service to the forests. Their hidden pathways through treetop halls of dappled sunlight and fluttering green. Their winter shelters clasped in a hundred maple palms.

I saw the vital, intrinsic worth of a squirrel as indistinguishable from the creature itself. A squirrel, any squirrel, felt like part of the overarching mystery and dignity of nature without distinguishing itself as an individual of particular merit.

I, on the other hand, had long felt that I had somehow started in a deficit and only immense work and achievement would bring my worth up to neutral, let alone the positive. A squirrel could just be a squirrel. I needed to be the hero of not just my story, but the collective story of the cultural values with which I was raised.

When I looked into my backyard, seeing a squirrel take an offered seed, my bruised mind trying to grasp any firm point in the weightless haze of my pain and uncertainty, I felt the tug of a wordless lesson in those creatures' surety of place and purpose. I couldn't conceive of a world where they were questioning if they deserved their food or their spot on the bough. I couldn't imagine that they questioned whether their lives were adding up to something worth having, that they were doubting past leaps or the quality of their dreys.

No, they seemed to be more *here* than I was, sitting in my dim kitchen watching out the window. I so rarely felt *here*. I was elsewhere. Assessing conversations I'd had years before. Measuring myself against old expectations and unkind imaginings. Trying to contextualize my life's story in a huge, abstract landscape of finances and relationships, career hopes and old hurts. I had to be elsewhere in order to stand apart and judge myself.

Shame.

Beyond my general sense of smothering sadness, the key emotion of those bleak hours of indecision was shame.

Shame isn't a creature of the here and now. It's a tether to old mistakes, failed expectations, and predicted futures full of further disappointment.

Shame is one way our human affinity for narrative becomes weaponized against us. With shame, we aren't just a character in a cohesive storyline, we are the villain. The moral lesson of the tale centers on ourselves as examples of wrongdoing. Traditional stories need an antagonist, and it is easy to slot ourselves into that role.

The simple truth that we are the ones crafting the storyscapes within our own minds ceases to matter, because we have long ago lost the belief that we have any agency in the telling. We begin to feel that our judgements aren't subjective—building harmful stories from our personal interpretations of events. No, we believe we are objective—passively observing things as they actually are.

This is the darker side of our ability to make meaning, and one of depression's key weapons for defending itself from treatment.

It's difficult to pursue healing when you are certain you don't deserve it, or when you feel confident that all your sadness is a reasonable response to a broken life, a flawed self, and a malicious world.

What do I need to heal from? Being realistic? Paying attention?

Beyond the illusion of objectivity, shame also shifts our focus away from the real time and place of our power. Our agency dwells in the present moment. The here and now is the place where we can actively exercise control. It becomes impossible to access our power to shape our lives and outlooks when shame is forever shoving us into memories of a painful past and twisted assumptions about tomorrow. Abstract narratives of failure and hopelessness keep our attention on spaces where we truly have no ability to effect change. Depression and shame force us to practice and rehearse powerlessness until it feels like a defining feature of who we are.

Watching the squirrels in the maple tree, in the deep winter of my pain, these realizations were taking shape. Yet, even when I clawed my way back to the present moment, shame was not toothless. With the identity-affirming structure of my job lost, one question arose again and again.

Who is it I'm trying to save?

A failure.

A lazy freeloader who is failing to appreciate even the immense privilege of having the time and space to seek healing.

A poor example of a man, overly sensitive, crying at nothing.

A misfit who was born into the wrong world.

A coward too ineffectual to cast aside a life that is clearly beyond repair.

None of these identities felt worth the effort of breaking free of pain and risking the disappointment of another failed reach for healing.

I saw a reflection of myself in the window. A pale, bearded wraith in worn-out sweatpants. Sunken eyes and a haunted expression. A person who looked like he had every reason to be sad and ashamed. The villain of the story.

I looked from my reflection back to the world beyond my window and felt disagreement.

Out there, the world of the heron and the squirrel, felt somehow incongruous with my rigid self-condemnation.

I spoke my argument up into the winter-bare trees that stitched the horizon to the sky above the gravestones.

"I don't deserve to feel content and it's pointless to try."

I was surprised to find that the trees did not agree.

"No," they answered.

"No."

"That kind of thinking may hold up in dim kitchens piled with dirty dishes, but not out here, not beneath this immense and ancient sky."

I tried again. "I'm too broken to fix. The world is too broken. It's too late. I'm not worth the effort."

The thoughts struck the landscape with an inharmonious *clang*, then clattered to the ground.

That wasn't the message of the heron.

That isn't the lesson of the squirrels.

I was missing something.

I felt it.

Something hiding in plain sight.

I was being crushed beneath the weight of a story that wasn't serving me, a story of a sad, lost man who disliked his job and felt that all his efforts were wasted. Yet, I knew there was more to my story than just this. I knew it. It seemed to have something to do with the feeling of kinship I had with the creatures outside my window. *What was that?*

If we can look at the living Earth and feel joy and love in the sight, can the essential truth of who we are really be beyond redemption? That feeling, that draw to the wild, it arises from two like things calling to one another. Reciprocal. A conversation. A moment of harmony. The pull of a family resemblance.

I sighed.

Hope, even a small, new hope like mine, can be a heavy weight to carry when you've grown unaccustomed to its heft.

A squirrel chittered in the sugar maple beyond my fence and a memory glinted green and gold deep down in the leaden waters of February.

It was a boy in a tree.

A boy who was right where he belonged.

❖

The tree my family called "Jarod's Tree" was a sugar maple.

A huge, old grandmother of a maple that dwarfed the neighboring trees. My childhood home was surrounded by patchwork woodlands, but they were young woods. The area had once been a vast apple orchard, and nearby street names commemorated that past. McIntosh Drive. Russet Lane. Golden Drive. Jonathan Lane.

I imagined that my tree once stood in those forgotten orchards, tall and untamed next to the rows of apple trees, its spreading canopy declaring a youth in open spaces without much competition for sunlight. When the apple trees faded into memory, then faded further still, and the cottonwoods, maples, walnuts, and oaks marched back in, my maple had quite the head start in the race skyward.

They called it my tree because of how often I sat in its branches.

Throughout my middle school years, my mother would shout toward the tree if she wanted me to come inside for lunch.

As I moved into my teens, many of my peers lost their enthusiasm for climbing trees.

I did not.

High school friends would drive along the wooded roadway and call up into the treetops, often enough receiving my reply from above. Day or night. Rain or shine. There was a fair chance I would be in my maple tree with grimed hands and searching eyes. I was an anxious, distracted, depressed teenager, and the place I found relief was the same when I was sixteen as when I was ten, nature as a vital and unnamed medicine I would one day regret forgetting.

I knew that tree.

It kept my secrets.

It was my fortress when a white-tailed buck was being loud and territorial on an autumn night, or when I felt bruised by the petty meanness of school politicking.

It listened when I pressed my forehead to the bark and thought my stories into the heartwood.

It even spoke back, in its own subtle ways.

I took special pride in how quickly I could pull myself into the lowest branches seven feet above the ground.

It had areas that I often equated to rooms in our house.

There was the living room, a bouquet of forking branches spreading from the widest part of the lower trunk. Here, I could host a friend or two with enough comfortable perches to go around. Higher up were the bedrooms, specific branches that were good for sitting up or laying down. Perches from which I could see into our kitchen window above the garage or study creatures that came to drink from the small stream that ran through our woods.

My favorite branch of all angled up from the trunk in just the right way to make a comfortable seat thirty feet above the ground with perfect footrest branches beneath. I would often straddle that branch and lay there like a big, dozing cat. Smelling the breeze. Watching insects come and go. Talking back to a surprised blue jay scolding me from a higher perch.

"Jay-Red! Jay-Red!"

"Oh, hush. I was here first."

It felt special and safe and secret. Even at ten, I recognized the magic of that place. There I was, an odd kid with an odd school career, simultaneously placed in the gifted program and diagnosed with learning difficulties ranging from minor dyslexia to ADHD, entranced by the quiet happenings in the branches of a sugar maple. Away from my peers. Away from the television. Away from my video games. Away from my books and toys.

In those days, I would have said the tree was "fun," but I don't think that's quite right.

The tree was medicine.

I remember, even then, recognizing the different sort of wakefulness I felt while alone in nature. I wasn't there to be distracted or engrossed in the way of my other favorite pastimes. Sequestered in those branches, I wasn't there to perform a function, not to be funny or smart or responsible. I was there just to be there, and I felt an immense connection to everything around me stemming from the special taskless essentiality of opening my senses to nature.

I was present.

It's amazing the number of mental hurts that are soothed by simply focusing our senses on our immediate place and moment.

When an ant crawled across the gray bark and a sunbeam made it shine, translucent as a drop of honey, I drank in the experience with all of my awareness. I felt no call to contextualize what the ant meant for my life. I wasn't considering how the ant upheld or rebuked my identity.

The ant had nothing to do with evaluating my worth as a son or student or friend. The ant didn't care if I was cool or smart or strange.

It was an ant walking on a maple branch, a common sight ten or ten million years ago.

While watching, I felt I was the ant—that our shared life and moment were more relevant than our individual differences.

I was the life I saw shining in the small, golden creature. I was the life I felt as sure as gravity in the tree beneath my hands. I was *there* in a way that didn't ask for further analysis, that didn't need to be justified or defended, that made abstractions like shame vanish like morning dew as the day warms.

That quiet treetop kindled an unspoken, uncomplicated wholeness in me that feels as vivid in my memory now as any memory I possess.

Too much of my past has been cleared away by pain and illness, but not that tree. Not that ant. Not the feel of that unknowable, healing magic coursing through the living wood.

Shame often arises from a judgement of our life's project as a whole, unified work. I was well acquainted with shame and sadness as a child, but I also had a friend in nature, a friend that constantly reminded me that our lives do not require formal interrogation. That we don't live as mathematical sums of events, successes and failures. Our lives exist in their realest form here, in this exact moment, and when we apply ourselves to fully witnessing the here and now of our strange and beautiful universe, shame becomes an illusion, just a trick of the light.

❖

I remembered my tree and who I was when I was in its branches. There was someone who deserved healing. There was someone who was not beyond redemption, who rejected a need for redemption.

Of course, my depression had responses for such thoughts. Depression always has a response.

You aren't that boy anymore.

That was before you were broken.

You still had a chance back then.

A million mistakes and squandered opportunities separates then from now. An ocean of time and failure.

What can one tree or one ant do in the face of real, justifiable despair and a bleak world?

Being present is nothing but a closer look at agony.

The only choice now is escape, is death.

What is it about depression that makes it such rich soil for cultivating shame? It's certainly tied to our cultural narratives and arbitrary societal barriers. It's also a function of becoming habituated to avoiding being present in the moment. But it's more than that. The character of depression itself is insidious, so that an examination of the concept becomes a kind of trap. I think it has something to do with the nature of mental illness, specifically the mental part.

When we suffer from an ingrown toenail or a toothache, we rarely jump to the conclusion that the affliction is evidence of our flawed personhood. That's because these aren't maladies of the mind. We like to imagine our bodies are just the machines that carry around the "real" essence of who we are—they are the ships we pilot, the soil in which we grow.

Bodies get sick. They age. They fail. That's intrinsic to their nature, and that reality is bound up in what we understand to be the human experience. The mind, on the other hand, is supposed to be the real us, the part of us that is not so crude or mundane as muscle and blood and bone. So, if the mind is sick, that's an indictment of our "who," not merely our "what." It means our identity is sick, not our physical substance.

This appears to make a kind of sense—that's the tricky part—except it's completely ridiculous.

Our brains are special, fascinating miracles of the natural world, but they are also organs of the body. They are flesh and water and electricity, and the fact that they are the seat of our thoughts does not negate their physicality or susceptibility to sickness. Depression is, without any doubt, an illness and should be thought of as such.

If a virus hijacks the machinery of your cells to manufacture more of itself, you wouldn't take ownership of the decision to produce viruses, would you? If you have a reaction to poison ivy, you wouldn't measure your self-worth by the itching of your skin, would you? Can you imagine considering chickenpox to be an indictment of your life or evidence that you are lesser than other people?

No.

So how is it that our self-worth and the value of our personhood becomes tied to the painful negative thoughts that depression sufferers neither willfully create nor invite into our lives? Depression is not a mirror held up to our identities, nor a yardstick with which to evaluate the quality of our essential selves.

It's the flu. It's a rash.

It's the emotional equivalent of a persistent headache.

Of course, the common cold doesn't typically make its victims wish for death, and depression often does. That's part of the problem. Beyond the very real pain of depression, emotional and physical, the shame aspect begins to make suicide seem like not just a better alternative, but a sensible and honorable rejection of a faulty self. It goes beyond

"I don't want this life" and layers on "this life is objectively trash and should be treated as such."

Recognizing depression in ourselves is already a difficult task because the disease has endless ways to hide within the thickets of our minds. For example, I've met depression suffers who, when asked about their mental health, reply with some variation of, "No, I don't have depression. My constant sadness is a reasonable reaction to a miserable life."

I would not, for a moment, suggest that there are no reasons for natural sadness, nor am I remotely qualified to offer diagnoses to others, but I do feel in a position to state that depression will always tell you that your sadness is something other than an illness, that treatment is irrelevant in your case. Depression is very convincing. It speaks with bold certainty, declaring that it's ridiculous to equate your reasonable, justifiable sadness with mental illness. The term "depression" is fine for other people, but you have *actual* problems that aren't "all in your head."

These deceptions are part of the nature of depression, shame being the key factor. We feel shame in admitting that our minds, the part of us we consider our real selves, can become ill. So, we fight to call our hopeless sadness anything but mental illness. We internalize the shame of societal stigma, of being a gray squirrel in a world of birdfeeder baffles, and we become the villain of our central story, a story we don't seem to control.

Depression and shame, along with the culture that links them, teach us it's better to be bad and damned than ill.

All of this shame and doubt can begin to make the key pathways of healing seem paradoxical, in the sense that our best tool to fight depression also remains the stronghold of the disease—our minds.

Despite all the illusions that depression can conjure, our brains are still our best ally in this fight. Yes, it's the seat of the disease, but you need to trust me on this most vital point: depression does not control all the territory of your mind.

It wants you to think it does.

It's lying.

Through the mechanism of shame, depression will tell you that the entirety of your mind has been coopted and rewired to produce only doubt and hopelessness and agony, but that just isn't true. Your brain is more complex, resilient, and expansive than that. Depression is an occupying force in the castle of our thoughts, but we each have a wide variety of secret passages, hidden rooms, and sliding bookcases from which to wage a guerrilla war to take back our skulls.

We may even discover an old, disused treetop refuge to bring us back to ourselves and our present moment.

❖

I stood at my backdoor and studied the branches of the sugar maple in the cemetery, wondering if I still had the aptitude for tree-climbing, wondering how long before someone called the police about a grown man, dressed like a toddler in a gray T-shirt and gray sweatpants, climbing a tree in a cemetery next to a rusty fence.

I smiled.

I was not brave enough for that stunt.

So, I summoned up my willpower to try out a different kind of stunt. A quiet one. I reached for the branches of a very uncomfortable thought.

What if it all works out in the end?

I squirmed under the weight of this thought, of what it meant for my shame and for my road ahead. Just the idea of *having* a road ahead was uncomfortable.

My sense of shame felt outraged.

It was obscene to allow myself a happy outcome.

How dare I even suggest a happy future?

It was too late.

The book was shut.

Even if it wasn't too late, I didn't deserve healing anyway.

I was too broken.

I was the villain of the story.

But the sugar maples were speaking from my past and present. Two trees. One in the nearby cemetery. One in a memory of my childhood home. Both connected to my present, to a tired man standing at a real and imagined doorway.

I dealt with depression as a child too, but I also had moments of happiness and peace. It could be done. Pleasure felt theoretical, but intellectually I knew it once existed for me. I had memories in treetops. I had moments of wonder peering under stones and lying beside muttering creeks to search for crayfish, their little plumes of disturbed silt billowing like underwater smoke as they shot backward from my searching fingers. I didn't need to earn those moments then, and I didn't need to earn them now.

What if it all works out in the end?

Sure, we all die and every life has its shares of tragedies, but that isn't the pushback I was feeling against allowing for the hope of things working out.

Sure, if we measure ourselves against a large enough cosmic scale, we will seem too miniscule to matter, but that wasn't the pushback I was feeling either.

The pushback was from my entrenched worldview that I was a miserable person, deserving of my misery, and the only reasonable, virtuous act I had left to me was to die. I was trash. I could do everyone a favor by taking myself out. That worldview, that identity, just could not sit with the notion that any other future might be achieved. Shame had become my reality, past, present, and future. I had learned to root against myself. How could anything "work out" in such a context?

We seldom admit the seductive comfort of hopelessness. It saves us from ambiguity. It has an answer for every question: "There's just no point. It's just not worth the effort."

Hope, on the other hand, is messy. If a good outcome is possible, then we have things to do. We must weather the possibility of happiness. We must hush shame and doubt and participate in the art of living because we are needed to usher good into the world, for others and ourselves.

I knew, when I was able to summon the raw honesty, I couldn't predict the future, so ...

What if it all works out in the end?

What if my portion of pain and sadness ahead was interspersed with happiness and peace? With worthwhile moments? What if there were still things I could learn or admire in the world? What if I could relearn how to nurture myself? What if past mistakes didn't actually cancel out future joys? What if building a life was a pleasant enough project even when I sometimes missed my mark? What if any simple pleasure could be the meaning of my entire life for as long as I could hold that meaning in my mind?

What if there, standing at the back door, letting the sugar maple of my present and the sugar maple of my past converge across a gulf of years could be the entire point of my life, could be enough to sustain me? What if each worthy moment was its own justification for existing, without being measured against forever, or earning me wealth or social capital? What if, moment to moment, I got to define what "working out" means?

What if it all works out in the end?

I rested my hand on the glass and sent my thoughts out toward the tree.

"Hello. I'm sorry I haven't introduced myself until now. I'm Jarod. We're neighbors."

It felt ridiculous.

It felt absurd and correct.

I sank to the floor and pressed my palms into my eyes.

Hopelessness is painful, but so is hope.

The pain of hopelessness is that there is no rest in sight from the pain and misery. The pain of hope is that happiness and peace are possible, and possibility opens new paths to exhaustion and disappointment.

Hope is, all by itself, a courageous act.

Hopelessness and shame both call for us to slam the door on ourselves, to close the case and end the questions, to avoid the painful "what if" imaginings that are the seeds of potential change.

There was the tree.

Here was the boy who once knew what the tree could be, what it could mean, how it could feel like family.

Between the two stood a fearful choice.

Could I reject shame and reach for the branches?

Could I grant myself grace and empathy?

I didn't know, but I felt certain that trying was the only path forward.

Chapter 4

RED-TAILED HAWK

THE FIRST TIME I THOUGHT I MIGHT LOVE Leslie was when she screeched at me like a red-tailed hawk.

Red-tails have a wonderful screech, loud and long and full of grit.

Their calls are often dubbed in for other raptors in TV and movies. For example, on Steven Colbert's old show *The Colbert Report*, the bald eagle scream in the opening credits was actually a red-tailed hawk. Bald eagles make chittering, cooing sounds that do not inspire thoughts of wild ferocity.

I met Leslie at Ohio University when I screwed up her first day of teaching.

We were both grad students who also taught low-level English courses.

I had misread the Freshman Composition class schedule, and when she arrived to her first-ever classroom full of students, I was already there introducing myself as the instructor.

She politely shooed me.

It wasn't the best first impression.

As we swapped places, one student said, "Oh, good, we get the pretty one."

Stay classy, freshman.

I got to know Leslie better in our graduate poetry workshops.

I disliked her immediately.

She never seemed to be taking any of our in-class discussions seriously. Early in our acquaintance, in Mark Halliday's class, we were workshopping a poem I had written about an ant-infestation as a metaphor for my deteriorating first marriage, and when my classmates handed me their notes on the piece, Leslie's feedback included a large doodle of a black ant eating a crumb.

It was ridiculous.

I had come to grad school as a nontraditional student, older than my peers, leaving my job at a real estate development business to pursue a life of the mind. I had failed as a student when I first went to college directly after high school, but when I returned in my twenties, I had a much different level of focus and commitment. Unlike my first brush with higher education, I knew why I wanted to be there.

When I reached my master's program, I took academia (and myself) very seriously. I wrote dense essays about John Milton's *Paradise Lost*. I attended conferences on pedagogy. I thought deep thoughts about deep thoughts, and I was eager to tell you about them. I used words like "epistemologies" before breakfast, without a hint of irony.

My mental health struggles, my insomnia, and my fear of fully committing myself lest I prove my best wasn't enough meant I missed class fairly often, but I was still on the path to being respectable. I was becoming a scholar. A writer. A professor. Somebody who left his small, rustbelt town to do smart things with smart people. I didn't have time for frivolous women and their frivolous ant doodles.

I was an idiot.

A pretentious, insecure, superficial idiot.

Neither Leslie nor her drawings were frivolous.

Eventually, I learned she doodled as a coping mechanism. Her ADHD and PTSD were held at bay through constant multitasking. The drawing wasn't her way to avoid listening. It allowed her to listen.

I learned she independently printed smart, funny, autobiographical comics and was a voracious reader of science fiction and fantasy.

I learned that her bones hurt when the weather changed, from over a decade of being thrown from horses.

She started riding when she was seven and began working at a stable that specialized in rescue animals near Detroit, Michigan, when she was thirteen.

At a department event at a professor's house, I found out that among all the grad students, only Leslie and I were comfortable building a campfire. Or at least we were the only ones eager and dorky enough to do it.

I learned that she was often up for a 2 a.m. walk on the edge of town or a spur-of-the-moment hour-drive to the nearest Waffle House in Lancaster.

She played video games better than I did, and when I mentioned my time dabbling in boxing, wrestling, and martial arts, I learned she had a brown belt in Judo.

She had harrowing stories of a rough childhood with alcoholic parents, and an endless appetite for playful whimsy.

Sometimes, I found that our humor didn't mesh. Much of my humor was still a holdover from high school. It was founded on a pattern of "we're being meaner than any actual person would ever be to another person ... get it?"

Leslie did not get it.

The premise was lost on her. She grew up around people who casually spouted the sort of cruelty that I considered comic hyperbole. If, as a child, you are told that you are "disgusting" or "unlovable," exaggerated insults may not tickle your funny bone.

Other times, our minds were bizarrely in sync.

One day, I was driving Leslie and two other friends to buy groceries (I was one of only a few grad students who had a car, so I often helped people with shopping), and Leslie was struggling to read a distant road sign.

"I can't quite read it," she said.

Apropos of nothing, I shouted, "Form of the hawk!"

If you find that phrase inscrutable, I don't blame you.

But somehow Leslie understood the odd workings of my brain, screeched like a red-tail, and redoubled her focus on the distant sign, now armed with the imaginary hawk-power of enhanced vision I had invoked on her behalf.

Uh-oh, I thought. *I think I might love this person.*

It was a complicated thought.

A fearful thought.

It was a threat of more change when I was so desperate for stability. Love was not part of my five-year plan.

Leslie says the first time she realized I was interested in her was when I casually explained how I was certain we couldn't be biologically related.

So poetic. So romantic.

But I had my reasons.

Part of why I misread the class schedule on Leslie's first day of teaching was that we both had the last name Anderson. So, one day I explained to her that I was actually an Anderson through adoption. My paternal great grandfather, Roland Anderson, Sr., acquired the Anderson surname when his birthmother, Laura Kelley, gave him to Anna Belle Teagarden Anderson and William Warren Anderson to raise, along with a cow to feed him, so the old family story goes. Leslie later commented, "I could only think of one reason you would tell me that in so much detail."

I'm sure I thought I was being subtle.

My first marriage had ended a handful of months earlier when I asked for a divorce, closing a decade-long relationship that began in high school. It was painful. It felt, at the time, that a full third of my life had ended in failure. Tara was my first love, and the fact that we wanted different things and had screaming fights several times a week didn't make the decision to part feel any less like a death in the family. It was the loss of a loved one. Even very sensible divorces can wound, can be a tragedy while also being the correct decision for everyone involved.

After the divorce, Tara moved back to central Ohio and I moved from a rented two-bedroom house into a 275-square-foot basement apartment tucked beneath a stairwell. It was a bed, a thrift store armchair, a tiny stove, and a closet/bathroom with a shower smaller than most refrigerators. It was full of spiders, silverfish, mice, and house centipedes. There were no level surfaces, straight walls, or right angles. It smelled like mildew and sour beer. I didn't have hot water until five months into my lease when Leslie insisted I call the landlord "again" to complain. Truthfully, I hadn't bothered calling before she prompted me. Hot water didn't seem worth mentioning until she wanted to shower at my place.

The apartment had issues, but it felt simple and knowable. It was exactly what I wanted after the murky complexity of my divorce. It was also a very short walk to my favorite hangout, Donkey Coffee, and about a dozen bars. These places were my actual living space. The apartment was just the hole I crawled into to sleep.

I was in mourning, and I wanted my life to match.

I wanted to be aggressively, belligerently single.

I wanted to make bad decisions.

I wanted to stay out late drinking and have nobody to wonder where I was.

I wanted to be a broke, eccentric scholar who wore misery like a badge of office.

I wanted people to think, "Wow, that guy is a wreck, he must be some kind of genius."

I wanted to be a dive bar monk.

Only I didn't actually want any of those things. Moreover, the life I had in mind doesn't exist, not in the way I envisioned. It's a childish, two-dimensional costume to shield against the vulnerability of earnestness. A caricature of a writer in a movie. A self-serious student in a bad play.

I was trying to escape into a romanticized shorthand of personhood that exists primarily to aggrandize mental and creative work as a kind of spiritual nobility. I wanted a clear path, even if it was a bad path, so long as it was easy to follow through the brambles of ambiguity, so long as it would silence the big questions about who I was now that I wasn't a husband, so long as it was easy to sum up at cocktail parties.

I mostly failed at being self-destructive.

I tried one-night stands and found that, for me, good sex involved emotional connection.

I found that being drunk or high did not soften the edges of pain; if anything the loss of my concentration softened my defenses.

Moreover, I had too many friends who loved me and wanted the best for me.

I failed at adopting the wretched, lovelorn scholar identity. It just didn't agree with my upbringing, my empathy, or my tendency toward silliness. Pretention is a solid way to hide from authenticity and doubt, but the price is high.

I still had too much humor and general distrust of authority to pour myself fully into career-driven scholarship or the power struggles of academia. I was part of a community of fascinating, like-minded nerds who turned their backs on practicality and, in defiance of self-interest and capitalist norms, moved to the hills of Appalachia to pursue advanced degrees in things like writing poetry or seventeenth-century British literature or Chaucer studies.

These were people who would demand I study French with them at tiny bakery cooperatives or carefully pick the perfect bottle of wine to pair with a hike up to the looming, castle-like former mental institution that frowns down on Athens from the ridges above campus.

People who cooked dinner with me or insisted I join their poker nights or showed up on my doorstep to fetch me for fiction readings or ukulele concerts or road trips to twenty-four-hour horror movie festivals in Philadelphia. People who wouldn't let me disappear into my tiny, buggy apartment beneath the earth.

That place and period were a beautiful, absurd, chaotic rebirth, not achieved through any merit of my own, but through the persistent force of my community. It was a pure and fleeting time before the reality of rejoining American society set in and most of us were swiftly reminded why, historically, humanities-focused liberal arts educations tended to be paired with generational wealth. Not the way it should be, but the way it is all the same.

It was the kind of context where a strange, beautiful poet could screech at me like a hawk and hip-toss me into unintentional love.

And, in one of those unlikely kindnesses of life, it turned out she loved me back.

I'm often embarrassed by the version of myself I see in the rainy, redbrick blur of my Athens memories. I see an arrogant, gullible, pretentious kid (approaching thirty) who uncritically swallowed academia's most prideful, most antiquated notions of itself. I see a person who wanted answers so badly, he wasn't picky about where he found them.

Leslie liked that guy despite his failings.

And I am grateful.

When we first began dating, I was planning to leave Ohio the next year for a PhD program in Missouri, but we decided that good things didn't always need to last forever in order to be worthwhile, so we gave it a shot anyway.

I am awfully glad we did.

I got to know her.

She is a creative powerhouse and a cheerful oddity.

Her upbringing was largely mean and humorless, Catholic schools and a harsh, unpredictable homelife, yet she somehow managed to transcend the forces pressing her to become petty, numb, and obedient.

She understands hurt the way a piano prodigy understands music. She makes empathy an artform. In her years working with horses at the animal rescue, she earned a reputation for getting through to the tough cases and lost causes.

A skeletal horse would arrive, found abandoned to starvation in a field in rural Michigan, days away from death, and after arriving at the barn would refuse to eat. Leslie would sit with them in a stall or a pasture and read, day after day. Just a calm, steady presence, stringing together hours of shared history with a suffering creature until something clicked and the horse would decide to take a chance on living.

Leslie was there as an ambassador of kindness, an undeniable example that the world was not all hurt, isolation, and abandonment. Perhaps they sensed in her someone who knew pain and loneliness.

She endured the mean horses.

Bites and kicks and being tossed into walls.

She soothed the neglected horses.

Vacant, shattered things impatient for an ending.

She understood how to be there for broken, hopeless animals.

Yes.

I know.

Here, we pause a moment.

I understand the parallel I'm approaching, and it's time to swerve aside.

I'm writing a story about my chronic mental illness and have now introduced my wife as an expert in healing wounded creatures.

So, let me get out in front of this and state outright—I'm not building to a metaphor. I am not a horse. Leslie is not my doctor, my keeper, or my deus ex machina.

She did not appear in my life as if by magic to rescue me. She didn't find me starving in an empty lot behind a Walmart near Lansing. That would be simplistic, dishonest, manic-pixie-dream garbage. Leslie's virtues are real, but neither romance nor friendship solves mental illness, nor can we shore up the cracks in our foundations with other people's lives.

The story of seeking mental health is rarely a story about saviors.

Teamwork, certainly. Saviors, no.

Leslie is, indeed, my hero and my partner, but she couldn't climb into my mind and fight my battles for me.

She is a constant reminder of not only the good in the world, but our capacity to overcome pain and hardship. She reminds me that we can break personal and generational cycles. She reminds me that it all starts with a choice, a rejection of inhabiting the spectator role in our own lives and stories.

Of course, change is always on the way to meet us, with or without our participation, but there is magic in intentional change, in the audacious act of inviting change and walking to meet it. Much like opening ourselves to love or friendship or fulfillment, courting positive change is the sort of magic that is tied to risk.

What could be riskier than pouring our efforts into a chance at something better?

What if it fails?

Leslie's customary answer is, "So what if I fail? I'll try again."

An intentional change makes us vulnerable to a vivid brand of disappointment. In the passive mode, we can hold tight to the idea that the future is happening to us. We abdicate responsibility. We don't steer the ship; we endure as passengers. Life does what it will without asking for our consent. The active mode saddles us with responsibility, with potential blame. If a change we invited sours, then both our fate and our judgement has failed us. We become both the victim and the victimizer, double the hurt and culpability of the passive approach.

And yet, I have come to understand that abdicating my agency is the greater threat. The numb, self-deceiving position that if we don't earnestly try we can't actually fail ourselves is not the defense it seems to be. With this mindset, change is the enemy and we won't be its accomplice in our inevitable misery. Except, of course, it isn't actually a decision between change or no change. It isn't actually a decision between choice and choicelessness.

We will act. We will influence our lives. If we do not exercise our agency, our willful choice, and actively seek the change we wish, we are choosing the status quo and, at some point, that uncomfortable truth must be acknowledged. Hope is fearful, but hopelessness does not protect us.

I was afraid to fall in love with Leslie.

I did it anyway.

Not because I was brave or wise or intentional.

It was because I couldn't help it.

I am happy to report, over a decade later, I do not regret it.

To feel deeply is dangerous.

To do anything else is a tragedy.

❖

In winter, wild places feel like being in a hushed house with a sleeper in the next room. A living stillness. It isn't quite loneliness. More a pronounced absence.

The red-tailed hawk is an exception. He will not be hushed.

The sugar maples told me that there was boy in me who deserved a chance at healing.

I often listened.

Leslie told me to go for walks.

I often went.

Walks were my early gateway to healing. Being in the woods connected me with a more resilient, more content version of myself, as well as grounded me in the present in a way that eased the shame and self-condemnation of my illness. Studies have found that walking itself, among other forms of exercise, may elevate mood. Moreover, it was vitally important that I was doing something, anything, for the express purpose of seeking wellness. It was a way to "put a verb in my sentence," as my mother is fond of saying.

Leslie would ask me if I was ready to find a therapist.

I wasn't.

She would ask if I would talk to my doctor about medication.

I wouldn't.

She would encourage me to go somewhere natural and walk in the thin, winter sunshine.

At first, I resented her encouragement to leave the house.

"Leave me alone."

"I just need more time."

"Nothing will help."

"Effort is wasted."

"Everything is hard and it won't do any good."

"I can see nature from here, can't I?"

But I loved her and I could see the worry in her eyes and the tension in her jaw. I had refused so many of her requests. I knew I must be profoundly difficult to live with, difficult to care for when I wouldn't participate in my own care.

So, I would go.

It hurt.

Doing anything hurts when depression is screaming at you to turn away from the pain of wakefulness. Any action, any thought, becomes like light and sound to a hangover.

I somehow managed to trust Leslie more than my pain.

She asked me to go and I went.

I cannot overstate the importance of this. I would do it. I would go. I was actively doing something in the name of healing. It was a milestone. It meant that, firstly, I was allowing for the possibility of feeling better and, secondly, I was literally moving toward that new possibility. I was accepting the premise that my choices and actions could affect my mind. Those first steps made all future steps possible.

Which is probably why Leslie needed to remind me to leave the house so often.

A walk outside sounds so simple, but it was potent, difficult-to-swallow medicine.

Change. Effort. Hope.

Terrifying words, each one a risk of further mistakes and regrets. Each a stair I could miss, sending me tumbling down and down.

In truth, as I left the house for my walks, I often intended to "cheat." I made little, secret, backroom deals with myself. *Yes*, I thought, *I will go to the park, but I will only sit in the parking lot for fifteen minutes.*

Once I was there, I would see Leslie's eyes in my mind and would stretch the deal a little.

Okay, I will get out of the car, but I'm only walking to that shelter house.

I would walk to the shelter house and find, to my surprise, that it wasn't so hard to be outside. It might even be a bit nice. So, I would chance walking further and further still.

Those odd, internal deals may seem ridiculous, but I've come to understand them as important tools. Ridiculous or not, any mental trick that broke through the wall of pain and got me to do more than nothing became a blessing. Whatever self-deception it entailed, whatever fantasies of minor rebellion against my agreements to go, the important part was that I summoned the effort and went. I was leaving my house. I was doing the work.

I was beginning to make a new kind of meaning for myself, a meaning that spoke in a low, reassuring voice, saying:

The herons are out there somewhere. The squirrels too. Bluebirds and chipmunks and turtles sleeping beneath frozen ponds. Cardinals and wild turkeys. Vultures kettling, turning in the sky like a great wheel. Remember the boy in the tree. Take him to go and see. He deserves to go and see. There's no wrong way to do this.

The depression and the deal-maker in me both rolled their eyes at such thinking. It felt corny. It also felt honest and nourishing. I learned to endure that paradox.

Sometimes, even when it's hard to honor our sense of hope, or the idea that we deserve comfort, we can still honor our sense of who we are. Identity is a potent form of meaning, less a thought and more a belief. My sense of who I am said I shouldn't argue against walking among the trees. It said that I'm a person who walks among trees. That thought isn't a function of what I deserve, it's a function of my belief about who I am.

I remember one walk marked with a bloodstain.

I was at a place called Gallant Woods, former farmland reforested and purposed to celebrate local flora and fauna.

That day looked like a postcard of Ohio February. Six inches of snow. So bright, even beneath thick clouds, I had to squint when I stepped from my car.

The place was empty.

Two hundred and thirty-one acres of old growth woods, patchwork meadows, young forest, and restored prairies and wetlands.

Empty.

My solitary blue car in the parking lot, the warm engine still clicking and settling, conspicuous mechanical sounds in all that quiet.

I walked to one of the trails, noting the rabbit tracks. The squirrel tracks. The flock of V-shaped white-tail hoofprints converging on their own marked paths through the trees.

Most times I visited that park, I was there to wander. That visit, I had a destination in mind.

A day earlier, I had watched a red-tail spot and dive on a gray squirrel, two fistfuls of daggers dropping from the marble sky, a bright cream morning shot through with dark oak-branch capillaries.

The hawk saw me watching as it settled on its kill. It fixed me with a copper stare until I moved along the path. I didn't want to startle it away. That would have felt like a waste of the squirrel's life.

So, on the next day's walk, I went to revisit the spot.

I found it easily, a hundred steps from my car.

A halo of blood in the snow. An imprint of feathers. Tufts of gray fur like bits of ash dancing in the chimney draft of a cold hearth. Part snow angel, part crime scene.

A red stain on the elegant blankness of winter.

I looked at the blood and thought of my own death.

Of course I did.

I so often thought of my own death.

This time seemed different.

If I was to be that bloodstain in the snow, something was off about how such an ending fit in with my view of suicide.

I didn't think of my self-inflicted death as natural, not part of any system of nature.

The suicide I so often imagined was a rejection of the world, not an aspect of its workings, not part of any fundamental cycle of life. My death was to be a rebuke of existence, a rebuke of the world and of myself.

This squirrel's death didn't feel like a rejection of the landscape. It wasn't an interruption. It was in conversation with the landscape. It was part of the low, soundless breath of the wild's winter sleep.

This death was nature.

I looked up into the branches as if the hawk would be waiting for my return, an old pulp story of a killer and the hardboiled detective pursuing him, a return to the murder scene and a reunion above a chalk outline.

There was nothing in the branches above me.

In the absence of the real thing, I interrogated my memory of hawks.

If a heron was slow consideration and the poetry of constructing meaning, what was the bird that stained this snow crimson?

Death?

No, a hawk didn't feel like death to me. Most all of the living nature I know relies on the generosity of death for sustenance and resources and the space to evolve. That doesn't define the hawk.

Here was a fresh, red mark on a white page, but it wasn't a blemish or an editor's pen.

I stood and pondered.

Somewhere, a scream defied the stillness.

If the hawk was not death, then perhaps it was a close relative.

Change.

All change is an ending.

Like death, an inevitability. An often unwelcome reminder of all we do not control.

An often unwelcome reminder of all we do control and wish we did not.

Change.

I didn't see a hawk on that walk, but I saw two on the drive home, each sitting on a utility pole surveying snowy roadsides. Most of the hawks I spot are not in the woods; they are perched on human-made objects watching human-shaped spaces. They are surprising silhouettes against the sky, interrupting the uniform symmetry of scalloped powerlines.

Across North America, red-tailed hawks inhabit deserts, scrublands, grasslands, and even tropical rainforests in Mexico. I usually see them along rural roadsides. They neither invited nor created any of the human contexts they haunt, but they seem to thrive in them all the same.

Change can be a monster we curse. It can be a goal we chase. Change can also mean adaptation, our ability to remake ourselves in response to pressures in our environment.

I had been considering suicide for years. Maybe death could be a kind of answer, but not the sort of death I had been simultaneously dreading and craving. Perhaps I could look toward a different image of death. Change. Adaptation. The end of one way of life and the beginning of another.

Afterall, how many times had I died already?

Twenty-seven-year-old me is gone from the world, and the echoes of him that remain are not quite him.

The same is true for seventeen-year-old me and seven-year-old me.

Those people no longer exist, and while I hear their footsteps in the attic, they are not breathing the air of this present moment.

Every memory is a ghost and the house they haunt is you.

I am the rooms and the footsteps on the empty staircase.

What could be more natural?

We change.

We end and begin again.

Our cells. Our contexts. Our mindsets.

What is suicide if not the hope for change, a hope born of change we feel confident we can seek and achieve ourselves, that we can attain through our actions. Yes, suicide is a grim wish, but it is also an acceptance of our own power over our lives. In that sense, a wish for death can be a first step to seeking a new and better life. Perhaps this version of my life could end without precluding future versions.

I thought about it.

Maybe not every suicide leaves a body or a splash of color in the snow.

Maybe, sometimes, we can kill the person we are without killing the people we will become.

Maybe we can live for millennia in the old forests of North America and then make room for the new versions of ourselves needed to thrive along roadsides and cornfields, to be newly alive in an unfamiliar landscape.

A hawk on a powerline isn't broken because he isn't high on a gnarled oak. He is different, but whole. Adapted and undiminished. Perhaps life and death isn't just a binary. Perhaps it can be an evolving conversation.

When I arrived home, Leslie was wearing her hawk necklace.

A carved red stone on a brown cord.

Maybe it's childish to find meaning in coincidence.

Maybe there are worse things to be than childish.

Back in our Athens days, she told me the significance of that necklace. In undergrad she'd had a particularly painful day in which she'd lost out on a scholarship she desperately wanted.

After hearing the news, she stepped outside and found a red-tail perched high on a tree above her. She looked up at the bird and felt like everything was as it should be. It didn't take away the disappointment, but it did suggest that maybe she was on the right path all the same.

Good news or bad news, if we can step outside and see a creature as magnificent as a hawk sitting in a nearby tree, then maybe everything is alright. Maybe, wherever we find ourselves next could be a worthy home. She wore that necklace often as a reminder that any one disappointment is not the final word on who we are or who we may become. If Leslie had received that scholarship, we never would have met. I used to laugh at the magical thinking behind talismans. I don't laugh anymore.

Some changes feel like a death.

But death often fosters new life.

❖

I love red-tailed hawks, but I did not love the meaning I found in them.

I hate change.

Not intellectually.

Not philosophically.

But in a core, fundamental sense, I dislike the very idea of change.

I'm embarrassed to admit this. I want to be too wise to hate change. It is, after all, inevitable. It's also the key to all growth and movement. It's the door through which learning and happiness enter our lives. It's the beautiful foundation of all choice and creation.

I hate it anyway. In the absence of wisdom, I will lean on honesty.

My hatred of change arises from a good ol' fashioned fear of the unknown. I don't control the future and I can't predict every outcome. All potential change is a window left ajar through which my worst anxieties can creep while I'm sleeping.

So I hate it.

It makes for an awkward predicament to both hate change and feel certain that your life has become unsustainable and intolerable. Two bad choices. A potentially negative shift or the devastating prospect of life as it is.

On the surface, this makes no sense. I was deeply suicidal, and depression had become a physical pain that sapped my energy and robbed every crumb of pleasure from my days. So, how could change be anything but welcome?

Even when we say our lives are unbearable, if we're still here, evidence suggests that we are, in fact, bearing it. We can't ignore this simple, empirical data. What if I make a change and my life shifts from figuratively unbearable to literally unbearable? I've heard suicide equated to a jump from the tenth story window of a burning building. As the fire creeps closer, as skin and lungs begin to burn, the fall becomes the lesser evil.

A change could be a rescue ladder.

It could also be gasoline spilled on the carpet at your feet.

Depression is very persuasive in its argument that life could always be worse.

It's easy, from the outside, to suggest a positive change to a person struggling with depression.

"Try yoga."

"Change your diet."

"Go for a walk."

"Talk to your doctor."

Easy for you to say. What if it doesn't work? What if I muster up the tremendous effort to defy my own brain, expend energy I don't have, and all I get is another failure? Another cutting truth of inadequacy handed to the brutal self-critic inside my head. That cost is real.

Yet, there is an upside to advice in the face of depression, especially advice from a loved one. It's true that most advice feels cheap when suffering from the pain of depression. It is also true that if we are working toward healing, we must acknowledge that depression is a liar. Worse, its lies are spoken in our own internal voice. So, a trusted outside viewpoint can become a vital tool. Outside viewpoints are, at least, outside the battleground. They are beyond the corrosive power of our mental illness.

When our own minds feel like liabilities, it can be soothing to remember that, thankfully, there are many other minds in the world.

We don't have to trust all problems to our own judgement and expertise.

There are other perspectives and life experiences available to us.

We don't have to endure all suffering alone.

Once we allow for the idea that depression is a liar, it makes sense to seek other opinions.

The shame of depression may tell us that our internal lives are too reprehensible to share with other people, but the potential embarrassment of letting others into our thoughts may be an easier hurdle to clear than enduring the full fear of making substantial changes on our own.

Every important step I have made in my own mental health journey has been an incremental step. Inches, not miles. Tiny bargains struck with my own doubt and willpower.

I will drive to the parking lot.

I will walk to the nearby shelter house.

I will try a bit further, just to the sycamore.

And letting others in can be a first, fierce act of hope that sparks chain reactions beyond what we dare to imagine.

I let Leslie in.

I couldn't act on all her requests or suggestions.

I couldn't simply adopt her positive outlook on my life or my essential self.

I couldn't borrow her courage or insight or understanding of healing and empathy.

But I could try to tell her what I was feeling.

I could stay in conversations even when I wanted out.

I could, in light of the meaning-making work I had done already, take her advice about going to the woods.

I could see her and the hawk she wore at her throat and work to imagine that change was the kind of death I actually needed, not a bloody death, but a winter-death, a pause before new growth, a way to a future where everything could be alright despite my pains and disappointments.

She invoked the power of the hawk and, imaginary or not, that power helped me move forward.

I could hear a hawk's scream from a distant roadside and understand that natural fierceness takes many forms. Sometimes, it's in the biological pull to adapt. Sometimes, it's as simple and difficult as going outside, as allowing for the possibility of meaningful change.

Chapter 5

WHITE-TAILED DEER

THROUGH MUCH OF MY CHILDHOOD, my mother fed the deer.

She wasn't supposed to. I'm certain it wasn't good for them. My dad hated how they trampled the grass and made the back hill look like a cow yard, not to mention how they would casually approach our cars when we drove down our long driveway through the trees. It was a nuisance how they would tuck their fawns away in the bushes by our front door for safekeeping and eat every leaf of our neighbor's hosta.

Mom fed them anyway.

Slices of white bread.

Dried corn from the feed store.

Apples cut into quarters.

I swear, one morning I saw her standing in the yard in her bathrobe, sipping coffee, handing her buttered toast to a doe that strolled up to greet her.

Eventually, she did stop feeding them. Mostly. But it was difficult.

White-tailed deer were not just animals to my mom. They were legends. They were childhood stories that somehow stepped into the waking world of her adulthood, right into her own yard to keep her company while she sat on the back porch.

My mom, Molly Kathleen Kelley Anderson, grew up in a tiny "shotgun" house in the east end of Newark, Ohio. The term "shotgun" meant you could fire buckshot through the open front door and have it exit the open backdoor without hitting anything. A small, straight, hallway of a house with ceilings and doorways so low mom's two half-brothers would need to duck several times to follow that shotgun blast from front to back.

What can I say about Newark, Ohio?

I grew up there.

I have reconnected with other friends from my childhood who moved away and many of us have stories of rough times, places, and people in Newark. When my good friend Tyler moved to Chicago and returned to his hometown for a visit, a local asked him if the big city was a frightening place to walk around.

Tyler responded, "I'm not afraid to walk at night in Chicago. I'm afraid to walk at night in Newark."

Tyler and I had guns pointed at us more than once in Newark. His girlfriend's car took a random bullet while she was driving around one night. We were both pallbearers when the opioid epidemic took our friend Rob. Heroin ended more than one of my friendships, one way or another. Sometimes, we secretly carried knives to class, and we enjoyed days off during bomb threats to our middle school and high school. We got drunk together in alleyways, in the woods, and on our friend Chad's porch, scheming and listening to Pantera. We would get bored and enjoy activities like going into the woods to see how big of a hole we could dig in an afternoon. We built forts and bonfires. We shot air rifles at most anything that moved, including each other.

I had a mostly peaceful, mostly privileged, upper-middleclass childhood, but my hometown, like many small Midwestern towns, could be a hard, vicious place.

Newark's character seemed rooted in history.

According to the Licking County records and archives website, Newark was known as "little Chicago" for a span of years in the nineteenth and twentieth centuries, owing in part to its sketchy reputation, importance to the railroad, and its many speakeasies, brothels, and unlicensed boxing matches.

One of the "fun" community history facts we learned in school was about when, after the local authorities refused to enforce prohibition laws, a deputy marshal named Carl Etherington was sent to serve warrants to speakeasies. He was eventually put in the Newark jail for his own protection. A mob broke down the jail door, took him from his cell, and beat and lynched him on our town square, near the current location of the county government building.

In fairness, my memories and young perceptions about my hometown are certainly colored by my tendency to imagine the worst, the enduring fear and paranoia that accompany my depression. Newark is also a lovely town full of interesting, creative people and some of the most fascinating indigenous earthworks in the world, but something of that "little Chicago" past seemed to linger even into my childhood in the 1980s and '90s. It was the sort of thing I didn't fully detect until I had lived elsewhere for quite some time and until I compared my childhood stories to my peers' in college and grad school.

Mom grew up on the east end of Newark in the 1950s and '60s.

The east end was the poor side of town.

For much of Mom's childhood, they used an outhouse. Indoor plumbing came later.

They heated the house with a coal furnace. Coal was delivered through a chute that led into their dirt-floored basement.

They didn't have a shower until Mom's father, Paul Howard Kelley, a diesel mechanic everyone called Hubby, finally installed one in the basement next to the sump pit, a gravel-filled hole where water could collect and disperse. Hubby would wash-up in the basement when he got home from work, and my mom's mom, Dorothy Louise Huff Kelley, would sit at the top of the stairs and chat with him while he scrubbed the grease and grime away.

At the time, my mom thought this was scandalous. Now she recognizes it was sweet.

She grew up with two half-brothers and a half-sister from Hubby's first marriage.

Lorena (Reeny), Duke, and Gary.

Reeny still lives in their childhood home on the east end.

Uncle Duke's actual name is Elva Dalton Kelley, a name he liked so much he decided to go by "Duke."

Duke and Gary were legendarily troublesome kids.

Once, Dorothy was getting Gary ready for bed and asked if he knew where his brother was. Gary didn't want to answer, but eventually admitted that he had buried Duke in the yard. Dorothy thought he was teasing. He wasn't. When they dug Duke out of the ground he was more than a little blue from lack of oxygen.

Other charming stories include the time Duke and Gary bribed my mom for her silence when they used her for BB gun target practice while she played on her tree swing. The price for her silence was that she got to shoot the gun too.

They were poor people in a poor area of a poor town, but they loved nature.

They often visited a place they called Eddyburg.

This place tickles me for several reasons.

First off, I can't find it on a map and Mom can't tell me how they got there, so it feels a bit like Narnia or Terabithia, a half-real dream world from her childhood. Hubby built a log cabin there. It was part vacation spot, part sacred site.

Hubby could drive, but he died when my mom was only twelve. Dorothy couldn't drive, but she would find folks who would take her and her young daughter out to the woods of Eddyburg. It was their escape. Their holy place. Their garden center and sometimes their pantry. It was the spot where they hunted for morel mushrooms when the mayapples looked right and the spring bumblebees were bouncing clumsily about sampling the fresh, spring undergrowth.

Dorothy dug up wildflowers from Eddyburg and transplanted them to her small yard in Newark. It wasn't common to find a yard in the east end that had Solomon's Seal, bluebells, trillium, and bloodroot, but Dorothy had them. She had such a lovely garden of wild blooms that the local Girl Scout troop would visit her yard to be tested on their knowledge and earn their wildflower identification badges.

They had a big cherry tree in the front yard and two apple trees in the back, their branches hung with feeders for birds and squirrels. My mom remembers standing at the foot of the cherry and asking the birds if she could come live with them once when Dorothy hid in the bushes to tease her. Dorothy's giggles gave her away.

I tend to think of her as Dorothy instead of Grandma because I never met her. She lives in stories. She died of ovarian cancer before my mom was twenty, many years before I was born, but she has always loomed large in my life. Since I was little, since my family understood the shape of my interests and the pull the wilderness had on my heart, I've been told that I would have adored her.

"You two would have been the best of friends," my mom says.

I have three old pictures of Dorothy climbing a tree. They must be from the early 1950s. She is a skinny lady in sepia tones wearing a white top and a long, flowery skirt. White ankle socks and chunky black shoes with heels that don't look suitable for tree climbing. Incongruous outfit aside, she looks very comfortable up in that tree.

I'm told the photos were taken by Hubby on their first date. Him on the ground. Her up in a tree.

I can only imagine what Hubby was thinking, but as they were eventually married and he chose to spend the film and take the pictures, I'm guessing he shares my opinion of a woman who takes to the treetops on a first date.

Yes. I imagine Dorothy and I would have been friends.

One quirk of my mom's childhood that sticks in my mind is a mural on the Kelley's living room wall above the couch, an image of several white-tailed deer emerging from the woods.

The mural was huge. Perhaps eight feet by four feet.

That mural surprises me. It feels like an excess that they wouldn't have made room for in the budget. Yet, it was there. A storybook picture of deer, their obsidian eyes studying the tiny living room of my mom's childhood.

Some folks of my mother's generation grew up in homes with massive crucifixes and other religious iconography dominating their living space.

Mom had white-tailed deer.

She had a mother who hand-fed squirrels and kept carefully tended shrines to wildflowers in a rough part of a rough town. She had the yearly repainting of the wren house that hung in the apple tree and pilgrimages to the wilds of mythic Eddyburg.

So, as an adult, my mom fed the deer and spoke soft words to them.

She asked pregnant does if they were getting enough to eat.

When a fawn was stashed among our bushes, she would say, "We'll keep an eye on the little one."

She told young bucks to behave and that they missed a spot rubbing away the velvet on their new antlers.

I watched her do these things and I would do them too.

She radiated the awe and reverence she felt for the deer, and I felt the magnetic power of homespun magic.

She showed me how I could be near the deer while out in the woods, how I could put them at ease by not looking directly at them, by avoiding squaring my shoulders toward them.

I learned to recognize how the deer would freeze to assess your movement, your intention, if you were a threat, and, if the mood beneath the trees was just right, they would flick their tails, swivel their ears away, and dismiss you as a benign part of the landscape.

If they did decide to accept your presence, something magical would happen.

❖

As I made an effort to reconnect with nature, I thought of my mother often.

I thought of stories of Dorothy and Eddyburg and the old wonders I had unconsciously dismissed as obsolete. I thought of my family history. Wildflower walks with my mom. Camping trips with my dad. Afternoons spent high in the maple branches.

These ideas and stories were a part of my life, but felt locked away in glass cases, inert museum displays within the halls of my memory. They were kept at arm's length by velvet ropes of time and pain. Especially pain. They were my past. I knew that. I owned that. But such gentle remembrances spoke in a whisper, while my present struggles shouted in my face.

Toward the end of that first winter after leaving my job at Ohio University, the mental pain reminded me of physical pain after a car accident. I've been a passenger on two occasions when the car I was riding in was totaled by a brutal collision. The days after, both times, a purple slash appeared across my torso where the seatbelt had stopped me from flying through the windshield. Despite not having any major, obvious injuries, everything hurt. Everything. Internal structures I had never felt before hurt in a way that threatened my understanding of the geography of my own body.

That winter I felt a similar aftermath, but the aches and pains were mental.

Of course, all pain is mental. Pain is a product of the mind, one more way that organ creates our reality. The mind is the only place our pain exists. The stimuli that cause the pain may be physical or emotional, but the result is still a product of our brains.

Depression often has a component of physical pain, and the science says that the same pathways in the brain that let us know we've been in a car accident may also interpret depression as a very real, very physical hurt. This has been my experience.

I was exhausted. I was injured. I had a hangover from years of trying to pretend away a disease that was inconvenient and inconsistent with the identity I wished to present to the world.

Beyond the pain, I was profoundly disappointed.

The disappointment and associated shame felt impossible to shake.

The longer I was away from traditional employment, the more my shame took on nuances. I was one of an impossibly fortunate few who had the option to step away from my working obligations in order to recover, and I felt neither happy nor relieved. I did not feel fortunate. I felt crushed. I felt that anytime I decided to get out of my chair someone was turning a dial to increase gravity's hold on me specifically.

During my master's program at OU, my special focus was on John Milton's poem *Paradise Lost*.

In the poem, Lucifer physically escapes hell and it just doesn't matter. It doesn't matter because, wherever he goes, he carries hell with him.

> *Me miserable! Which way shall I fly*
> *Infinite wrath, and infinite despair?*
> *Which way I fly is Hell; myself am Hell;*
> *And, in the lowest deep, a lower deep*
> *Still threatening to devour me opens wide,*
> *To which the Hell I suffer seems a Heaven.*
> *–Paradise Lost*, Book IV

After years of study, Lucifer's words are etched on my memory. His suffering and regret cease to be a matter of place. They become a matter of identity, a defining characteristic of who and what he has become. He can no more escape hell than he can escape his own perspective. And, somehow, despite his legendary misery, he must allow that it could always get worse.

This idea crept in to haunt me.

Of course, I'm not a mythic hero or villain, but depression has a way of luring us into thinking in terms of grand absolutes.

Capital F failure. Capital S suffering. Capital D despair.

Depression does not feel mundane. It feels epic. It feels like, and indeed is, a matter of life and death.

I do not simply suffer. A verb. My life and myself have become suffering. A noun. A static state. Not an unpleasant day or an unpleasant

moment. No. Depression convinced me that I had become a personifica-
tion of suffering and, moreover, suffering as the correct and reasonable
result of a failed life and a flawed mind. A just punishment. A sensible
and inescapable hell.

That is the insidious nature of how depression works on a mind.
We make sense of the world through stories, often stories of cause and
effect. If we hurt, if we suffer, then what is the story behind the cause
of that hurt?

When we don't find a clear narrative behind why we constantly feel
misery, we may switch from an external "what cause" to an internal
"who cause."

The problem is not what I did. The problem is who I am.

We make our very real pain part of a fictional meta narrative.

For me, it was, "If work wasn't the problem, then it confirms what
I should have known all along. I am the problem. I don't just feel the
pain of a bad situation. I am the cause of the pain. There's no escape
from that."

A grand story of a broken life. A mythic tale of gods and devils. Not a
disease to be treated, not symptoms to manage, but a parable of woe, a
foul and broken mind locked inside a hateful cage of bone.

These were the lofty thoughts of a sweatpants-clad philosopher star-
ing at dirty laundry in a dark bedroom, the stuff of epic poetry. It's an
easy trap for me to fall into. As a human and a book-nerd, I need there to
be a cohesive story behind the pain. I need it all to make narrative sense.

I was someone who, in trying to understand the suffering of depression and finding no obvious cause at hand, no memory of a car crash
the day before, no bruises purpling my chest like a sunset, turned to the
big picture, created a story in which my entire life was a car crash and
there would be no end to the resulting pains.

We evolved to understand and assess danger, to learn from our pain, and when the pain seems to emanate from our entire lives, it becomes tempting to dismiss our entire lives as the danger, as the predator prowling the tall grass.

Shame is a typical self-defense strategy employed by depression, and it continually drove me back to this image of myself as a villain or devil. Rightly punished. Rightly suffering. The natural outcome of an evil existence.

When depression tells us we are broken, that we are correctly miserable because either we, or our world, is beyond redemption, we are imagining a point of comparison. This was a key realization for me, beyond the idea that mental illness is a disease. (I still might deserve a disease, after all.) If I had failed, the question was ... failed what? Failed compared to whom? The idea that this was somehow not the correct version of myself meant I must be imagining the version I had fallen short of becoming.

I was judging myself against a hypothetical ideal version of me.

The version I should be.

The version I must become (and cannot).

The distracting, destroying act of endlessly measuring myself, my life, and my world against what it "should be" made progress feel not just impossible, but irrelevant.

What's the point of working on my happiness when I am not even the version of me who should be here, the version who deserves happiness? This isn't the version of my life I should have. This isn't the version of the world that should exist. So, why bother to engage with it at all?

Working on this life, as it is, would be a tacit admission that this is the real me, my real life, my real world. Even to begin such an endeavor would necessitate meeting myself where I currently am. It would mean working with this version of myself, not mourning or rejecting the very premise of who I am now, but rather deciding that this version is worth understanding. Deciding that I, as I am, without caveats or edits, am worthy of the effort of healing and cultivation.

We can't intentionally change something we refuse to accept and perceive honestly.

Later, when I began learning about cognitive behavioral therapy, I would encounter the tongue-in-cheek phrase, "Don't should on yourself." Psychologist Albert Ellis called it "musterbation," a constant preoccupation with what you should or must do/be.

Mental health professionals do love their corny wordplay.

I equate the grasping paralysis of these ideas with wanting to get in shape before you go to the gym. This version of me isn't in good enough shape to begin exercising. This isn't the version I should be, so I must become that correct version of myself before I can begin to do anything. The already in shape version of me, he will go to the gym. My actual self doesn't deserve anything but scorn and regret. I don't owe this version of me any work, any care, any acknowledgement beyond persistent shame.

The problem is that a constant preoccupation with the "shoulds" and "musts" that remind us we are not living up to the standards of our imagined, idealized selves will stop us from loving and caring for the versions of ourselves that are actually here. We tumble into the mire of honoring hypothetical versions of us who live comfortably within an untouchable abstraction. We honor them to the detriment of our actual selves living here in this real, messy, challenging, sweet world.

These idealized versions of us have the privilege of not experiencing unexpected bills or sudden illness or any of the countless other difficulties that dirty the hands and faces of the real us. The "should/must" idealized versions of us enjoy all the honor and none of the work. All the praise and none of the challenge. No wonder we can't measure up to them. They are pure, hollow victory. We are so much more.

Day after day, walk after difficult walk, I was becoming, once more, a student of nature.

Nature, like you, like me, is a beautiful mess.

It is struggle and mud and sudden downpours.

Often, it is apparent chaos masking subtle harmony. It is hidden magic. It is cycles of decay and growth, dormancy and renewal.

I couldn't help but notice that nature, as I was coming to understand its character, had no use for the idealized, hypothetical version of me. That version was too efficient to wander the woods and make space for unsought discoveries. That version didn't have soil under his nails and creek-water soaking his socks. He was off somewhere being productive, being successful as defined by someone with more authority than me.

He wasn't real.

He wasn't missing from my life.

He wasn't on his way here.

He never had been.

And I was finally starting to come to terms with the fact that I didn't much like him anyway. I certainly didn't need his permission or approval to be whole, to acknowledge the real me, exactly as I am.

My existence wasn't broken.

My old job might never have been my key problem—chasing after a fictional, idealized version of myself was. When I was quiet and attentive beneath the trees, I began to understand that my traditional view of "success" hadn't been mine at all. I didn't make it. It didn't serve me. It brought me neither peace nor pleasure.

I felt in my marrow that it was time to envision a new kind of success.

It was time to make room for myself, the flesh and blood version of me who stumbles and hurts and endures.

It was time for new lessons and new teachers.

It was time to reunite with nature, and that nature included my real and present self.

❖

I had been crying in the car.

Again.

Clenching my jaw.

Scowling into the rearview mirror.

Exhausted by the cycle of pain and numbness.

Exhausted by the prospect of breaking it.

I was parked at Gallant Woods, trying to summon the energy to begin the walk I had promised Leslie I'd take when I left the house. I made a habit of promising. Often enough, the promise was the only reason I found the energy to act.

I sat, listening to Christy Moore's cover of "Beeswing" and searching my heart for a way to admire the sodden, ragged landscape. Wet with snowmelt. Brown and battered by winter.

It was early March and, while I knew spring would arrive soon, it certainly hadn't arrived yet.

Ohio gray.

How very tired I was of Ohio gray on that colorless afternoon.

But I had promised, so I shut off the car and instantly missed the heat and sound, the beautiful sensory distractions that were a shelter from my own punishing thoughts.

I got out and headed for the path.

I wrinkled my nose at the smell of thaw. Decomposers and detritivores getting to work as tiny shards of shrinking ice drifted in puddles the color of cardboard.

I was tired of the crushed-gravel paths with their cheerful joggers and senior couples trading their small talk about last frosts and garden planning. Wishing to suffer an intentionally poor decision, I made for the shoe-swallowing muck of the deeper woods. No joggers. No casual walkers. Still too cold for mosquitoes and surprise spiderwebs stuck in my eyelashes.

I took a lesser-used side trail that led over fallen trees and across a wide, often-dry streambed with big stones spaced for a crossing when the water was flowing.

It felt, as is so often the case when I ignore depression's advice, better than I expected.

The thick leaf litter and the uneven terrain meant I could avoid the softest ground if I thought about my path, and carefully choosing where to place each step meant I couldn't ruminate on damning abstractions. I felt the familiar relief of nature unfurrowing my brow. Each step brought me back to myself, back to the present moment.

I won't say time in nature heals depression. For me, the outdoors changes sadness from a pain to be endured to a state to be experienced. The sadness and depression are still there. But in the context of living, growing things under an open sky, pain is simplified. Not a wound. A tile in the mosaic.

Even in leafless winter, the tangle of branches and underbrush, especially in younger woods, can be obscuring.

I didn't see the deer until I was forty feet from them.

I rounded a big, fallen cottonwood tree, its roots torn and jutting skyward in branching points like a jagged, out-of-place transplant from a coral reef.

I picked my way around the crescent hole the tree had left in its fall and I saw three soft brown heads turn in unison to study me. Black eyes. Black snouts. White throats.

White-tailed deer, fluffy in their winter coats, stone-still with all senses trained on me.

Having so much lean muscle tensed on your behalf is an almost physical sensation, the air pulled taught with potential energy.

I knew those deer could crash through the brush and be gone in an instant.

Deer may not have the endurance of a human or the speed of a horse, but they can move with rapidity through thick woods in a way that appears supernatural, shadows untouched by briar or branch. I've tried to sprint through brushy woodland. It isn't as easy as they make it look.

I changed my posture and averted my eyes, dropping out of a mindset of forward motion and into a mindset of casual curiosity. I turned to a nearby maple, skinny with youth and lack of sunlight. I leaned in to look at its lovely gray skin, at the lichen tattooed there.

I wasn't watching the deer, but the tension was still there.

I sniffed the tree. I studied my hands. I checked my shoelaces.

Slow, steady, unconcerned movements.

These were my attempt at speaking the language of deer, of saying with my imperfect understanding of their speech, "I'm not hungry, and I'm not a predator. Not today. I have no business with you."

In my peripheral vision, I saw an ear flick toward some distant sound. A good sign. Perhaps my message was received.

I turned my back fully on the three deer and carefully took a step away.

No sudden sound.

No explosion of movement.

I very carefully sat down in the wet leaves, choosing an angle at which I could just see them at the edge of my vision.

They didn't like this.

One took an unbelievably silent leap and put another fifteen feet between us. The other two trotted along, following suit.

They paused, considering me again, now recontextualized by the added distance.

Deer are incredible experts at fleeing. I can't fathom how that extra fifteen feet changed the math of the situation for them, but it did. Maybe it was just enough extra lead that their instincts told them there was no way I could get to my feet and close the distance before they could vanish.

They didn't vanish. They just studied me again.

I picked up a clump of wet leaves and began considering them one at a time. A threadbare yellow maple. A brown oak. I discarded some and held onto others, fanning my favorites in my hand like playing cards. I brought them to my nose and breathed deeply. There were seasons in that smell. There was a conversation between life and death, spring and autumn.

Another flick of the ears. Another. Another.

A snowy tail rose and fell.

Then I felt it.

The tension in the air melted away like a sigh, like an unclenching fist.

The deer shrugged out of flight mode and began a slower wander deeper into the trees, probably looking for another place to bed down for an afternoon nap after being disturbed.

They glanced back at me now and then, but they weren't hurrying.

A nuthatch landed on a tree across from me, closer than I would have expected. It glanced at me, then continued its hopping hunt around the trunk.

That tiny bird was a note slid across a broad gulf of years.

A message.

Not about something I had forgotten. Not really. I had not forgotten.

It was more a spotlight on a thing that had faded from importance in my mind. Not lost. Overlooked. Taken for granted. A toy cast to the edge of the garden, paint fading in the sunlight.

When the deer accepted your presence, something magical happened. That nuthatch reminded me exactly what that meant.

My mom told me about this phenomenon long before I witnessed it myself.

When the deer decide you aren't a threat, other animals will follow suit. Mom would hunt for mushrooms out in the woods near our home, and the deer that knew her well would approach to see if she had a treat for them. When they did, the birds, squirrels, and chipmunks would all become less shy. Cardinals landing within easy reach and chipmunks chasing each other two steps away.

A stubbornly analytical part of me wants to interrogate causation versus correlation here.

It may be that the same behavior that puts deer at ease puts other wildlife at ease too. That feels sensible. Yet, another part of me likes the idea that other creatures rely on the deer's judgement. I like the idea that, as masters of spotting and dodging danger, the deer get to decide for their community what counts as a worthy threat in the woods. Deer have the final word on danger.

The nuthatch circled the tree like it was a spiral staircase. Up and up. The deer moved on. I sat and felt the cold and wet soaking through my jeans.

I stood.

The deer had no context for me or my life. They had no narrative sense of me as a success or a failure, a hero or a devil. They weren't measuring me against who I was or who I could become.

They just had me.

Exactly as I am.

In the moment they observed me.

I put my arm around the maple and thought.

Would it really be so bad to see myself clearly and accept what I saw? A deeply depressed man struggling to keep his head above water. A person who made mistakes, a person who assumed that his working life was the primary source of all his pain and then learned different.

Pain.

Depression.

Illness.

Mistakes.

Were these unforgivable sins? Hell-worthy sins? Were these such stark violations of who I "should be" that I couldn't acknowledge them long enough to seek help? To move forward? To extend myself honest consideration and care?

No.

I was so tired of letting lies and hypotheticals chase me through the briar bushes, "musts" and "shoulds" hunting in packs.

Standing in those bare winter woods, hugging a tree to my side, feeling the deer emerge into an inner-clearing of my mind, I felt a shift in seasons.

I felt it as sure as I felt the tension melt away after the deer had judged me.

It was time to stop warring with the simple facts of my life and my struggles.

It was time to stop considering my pain to be a hateful secret, too shameful to speak aloud in the wider world.

It was time to begin the work of meeting myself where I stood, of seeing myself clearly and without constant, reflexive condemnation.

It was time to seek help.

It was time for spring.

SPRING

AN OHIO SPRING IS A GREAT, VIBRANT FORCE teetering on the gray pinnacle of winter.

At first, just a glimmer. A breath against the cheek. A hint of green along the highway.

Pebbles presaging the rockslide.

Then, the tipping point, and spring comes tumbling down in a gentle calamity of warmth and color.

The waters rush and foam.

Leaf and bloom renew.

Life spills forth in a sudden flood that makes the lately gone cold seem distant in defiance of logic and memory.

A song. A homecoming. A stirring in the branches.

An invitation to come and join the old ritual of growth and change, of welcome and reunion.

After the long winter, I was ready to seek help, but those early steps were as much about sitting with my fear as embracing transformation.

Winter threatens with lack. Spring with possibility.

Chapter 6

FORSYTHIA

FORSYTHIA IS A BEAUTIFUL MONSTER.

Resilient, invasive, and popular for its easy cultivation and showy, yellow spring blooms.

Growing up, I saw forsythia swallow a hillside and my mother's garden. My parents weren't terribly interested in or knowledgeable about gardening and forsythia betrayed them. They planted it as a divide between a sloping rock garden and the woods, but forsythia isn't the plant for respecting and maintaining boundaries. It leaned over the garden and rooted where its drooping branches touched the mulch. Leaned and rooted. Again and again. Until the hill, the mulch, and the carefully placed stones, dotted with the shell fossils common to the Ohio Valley, were overwritten.

Forsythia didn't guard against the woods; it conspired with them.

It seemed to work for the wild, acting as scout and outrider, constantly retaking tended, domestic spaces for the great green tangle beyond our yard. It claimed new terrain for the mantises, the wrens, and the orb-weavers.

I have an unusually vivid memory of the first time I was introduced to forsythia. I must have been around eight years old. We had just moved into the home my parents still occupy. A gray Cape Cod–style house on a couple acres of woods on the west side of Newark, Ohio.

Forsythia arrived in my paternal grandfather's fist, a bundle of short, leafy sticks.

I was immediately intrigued.

He had brought them for my parents to plant.

"Plant?" I asked. "There are no roots?"

Grandpa explained.

"It doesn't matter to forsythia. You just stab them into the ground."

It felt like a magic trick.

In the spring of my thirty-eighth year, the hardiness of forsythia felt much less magical, unless perhaps as an arcane curse. It felt adversarial.

April had arrived and the forsythia in my yard were fountains of sunny, yellow petals.

Leslie and I had recently moved to our house in Delaware, Ohio, with a mortgage based partly on the strength of my director-level job at Ohio University, a job I had just left behind.

Spring felt different that year.

All winter I had been tugging at the skirts of nature, asking in a small voice for its guidance as it stood dressed in sober grays and browns.

Now it was changing.

The sky was different.

The smell beneath the trees was different.

My own backyard was different, and I wanted to meet it with my senses open and my mind present.

I am not an expert gardener and never have been, but something about the landscaping of our new house seemed discordant to me.

Fast-growing blue spruces planted beneath powerlines to block the view of the cemetery.

A half dozen forsythia bushes.

A smattering of hosta.

It felt like landscaping meant for a quick, cheap return on investment. Inexpensive, nonnative plantings that would mature soon, but were poor long-term choices for the space. Perhaps a landscape planted a few years prior by someone who knew they wanted to sell their house.

I was trying to make friends with a new identity that spring.

It was still taking shape, still fuzzy and undefined, but it coalesced around two central themes.

A nature enthusiast and a mentally ill person seeking healing.

These were the two ideas I wanted to make peace with, to cultivate in the rocky soil of my self-image.

I looked out my back window and felt that tug of identity, the same one that plotted with Leslie to coax me into the woods. Except this time it said, *Shouldn't a nature enthusiast have a more active hand in his own backyard?*

The old, familiar "should" thoughts were rising up to tear me down.

I recalled the stories of my maternal grandmother, Dorothy, of her little garden where Girl Scouts studied Ohio wildflowers. Then I looked to the bland, green rectangle behind my home, the line of spruce and forsythia meant to hide a perceived eyesore, not to invite nature into my living space.

It was spring and, while I wasn't a gardener, I knew what spring meant for gardeners. It meant it was time to get to work. I had neither energy nor expertise, but I wanted to get to work too, in whatever way I could. Winter had taught me that doing a little was a world away from doing nothing.

A lover of local nature would not have an aggressive, invasive species like forsythia dominating his landscaping. That felt like an obvious, inflexible truth and a clear place to start.

I had some pruners and two shovels in my shed.

Those tools seemed like more than enough to get me started.

I would remove one of the large forsythias. That felt like a humble first goal.

Knowing what I know now, it is difficult for me to imagine a less humble first goal.

I was picking a fight well above my weight class.

❖

The winter had prepared me to seek help.

I was not happy about it, but I was prepared to do it. For Leslie. For the sake of stubborn hope. For the sake of a new remembrance of my past self, small and distant, sitting high in the branches of a maple within the twilight gloaming of my past.

The first step was my doctor.

This felt like an easier, more incremental move than seeking a therapist.

"Normal" people go to the doctor, I reasoned.

That last sentence makes me squirm now, but it was part of my thinking. Sitting in the waiting room of a family medicine doctor, nobody would glance over at me and know that something was wrong with my mind. They might think I was there for a more respectable disease or injury.

Dr. Bartman practiced family medicine through Ohio State University's healthcare system. I made an appointment at her small, brick offices in Worthington, Ohio, a thirty-minute drive from my new home. It was a difficult step eased by the fact that the first available appointment was several weeks away. I couldn't manage much specificity in the online scheduling form. I booked an annual wellness visit, but I listed "depression" as a primary concern.

Even that small step was terrifying to me.

I knew that discussing depression would likely mean trying medication. I didn't know what else it would mean. Brutal, invasive questions? The threat of intervention by law enforcement if I said the wrong and truthful things about suicide or somehow proved myself a danger to society?

One odd benefit of having been unemployed for a stretch of months was that I was falling out of practice with the endless excuses that had dominated my working life. Excuses to miss a meeting. Excuses to take a sick day. Excuses to come in late or leave early. Intricate webs of lies and half-truths to keep my gasping mouth just above the waterline of professional respectability.

But that winter I hadn't needed excuses.

No one was asking me to do much of anything.

Yes, Leslie made gentle requests of me, but even when I refused, I told her the truth about why I was refusing. No fabricated excuses. I was starting to live in a way that was honest with myself and others, a way that felt novel after employing constant coping mechanisms to hammer the square peg of my mind into the round hole of school and work.

But now?

I was an employee of the mystery of myself.

I was a student of my own mind and moments.

While I struggled to appreciate the rare gift of having space and time for healing, I felt that gift changing me all the same.

The day came, and I went to my appointment.

I noticed a stack of unfamiliar pages on the clipboard they handed me in the waiting room, a detailed questionnaire about my quality of life.

Do I think of harming myself?

Do I feel safe at home?

Rate the truth of this statement: "I find no pleasure in activities I once enjoyed."

I tried to be honest.

Even the sugarcoated version of my honesty made me feel like a fraud.

Do I really have it *that* bad?

Who am I to be here, complaining? There are people in this waiting room with real problems. Cancer. Diabetes. Broken bones. Why am I taking up space and resources better used for more worthy, more legitimate ailments (and people)?

Beyond the guilt, how forthcoming could I really be about my situation?

I can't tell them I consider suicide constantly. Can I? Didn't people get taken away for saying such things?

They will lock me up and take my belt and shoelaces. The stain of such a thing, if I survived it, might color my life forever. Was it too late to just stand up and leave the office?

I just want to be left alone.

Why am I here?

Why did I agree to this?

But I knew the answer to those two questions, and I knew I couldn't tell Leslie I left my appointment or lie that I stayed, so I studied each form on the clipboard and tried to walk a thin line between honesty and careful understatement.

I did my best.

They took me back to a little exam room and had me sit on the edge of the table, the crinkly white paper giving me the impression that my presence there was somehow disposable and wasteful, fast-food in a wrapper. I was ruining something just by sitting down.

The nurse took my vitals then left me to my thoughts.

I waited, tension drying my mouth.

Eventually, a staccato knock and in came Dr. Bartman, a businesslike woman with glasses and short, salt-and-pepper hair.

I thought I had undersold my pain in a prudent, responsible way, but Dr. Bartman intuited the severity of my depression and the real reason I had scheduled the appointment. I felt silly and transparent.

When we began speaking, the first thing she told me was that my case was hardly unusual and that she had discussed similar symptoms a dozen times that week.

It was helpful to hear my struggles framed as typical, but my attention was divided between her words and holding a neutral expression to keep from crying.

Moreover, my depression is a blackbelt judo master at redirecting positive energy into negativity.

Ah, so this is common? Then perhaps it's wrong to call it an illness? Perhaps the world is as inherently miserable as I suspect?

Fear surrounded the entire conversation.

Fear of the unknown.

Fear of losing my agency, my freedom.

Fear of saying the wrong thing and confirming my suspicion that I had no business living among other people.

In retrospect, I don't know why I didn't do research ahead of time about what sorts of self-harm statements would trigger more invasive interventions, but I recognized the relevant questions when they arrived.

"Do you have the means to kill yourself?"

"Yes. I suppose so."

I had weapons, including firearms. I didn't hunt, not for many years, but I owned several handguns and rifles kept for defense and trips to the shooting range to stay in practice. Although, honestly, I didn't need them and wouldn't think to use them for suicide. Death by gun seemed unnecessarily messy and rude.

I had jiu-jitsu and other martial arts training that told me that there are many painless, tidy ways to cut off blood flow to my brain without even interrupting my breathing or marking my skin. I had been rendered unconscious more than once while sparring and grappling. I knew of multiple, gentle methods to fade from reality, and it was not difficult to imagine ways to apply the same techniques to myself with common tools and materials. No, a bullet would not be my chosen exit from the world. I wasn't interested in making a dramatic statement or spilling my blood.

Truthfully, I had acquired a variety of knowledge that assured me that life wasn't terribly complicated to coax out of a body, not if that was your earnest goal.

I, of course, explained none of this to Dr. Bartman.

Not a word of it.

"Do you have a plan to kill yourself?"

"No."

I answered before giving this one much consideration. This felt like the tripwire question, the one that would have me sleeping somewhere monitored by medical professionals.

Upon reflection, "no" was probably the honest answer.

It's true that I had many ideas of how to accomplish suicide, but none I would call "a plan."

I was not planning to kill myself.

It feels odd to articulate it in this way, but I couldn't actually do it while being with Leslie because it would be unkind.

She would blame herself.

I didn't think she needed me. In fact, I often thought I was dead-weight, an albatross lashed around her neck.

But I knew she loved me and wanted me in her life.

It would be inconsiderate to disappear from her world.

A plan to die would need to be a plan to mitigate the collateral damage to her and others who cared about my wellbeing.

This has been a longstanding facet of my suicidal thoughts. In fact, one warning sign I felt keenly just before I left my job at Ohio University was a feeling that I was beginning to sabotage my relationships. I'm not sure where I would choose to end my life, but I am sure I would be utterly alone and likely somewhere that nature could reclaim my body without all the associated human fuss of embalming or cremation. Coffins and urns. If I could, I would choose to meet the plants and animals who would carry this body back to the great, cyclical oneness of our living universe.

And I would do it unobserved.

I would not die to make a point.

I would die because I was finished making points, finished enduring my own perspective.

Afterall, the idea would be to stop being "me" altogether, so I hardly needed my death to be a statement.

One of the things I've often thought about in the context of suicide is that it isn't just an escape from the pain of sensation. It's an escape from the pain of a human viewpoint, of always having to be contextualized though stories and identity and history and endless comparisons of relative merits and flaws. The pain of obligation and expectation.

Part of the pleasant allure of death for me, both inside and outside the conversation of depression, is a return to a kind of existence without the need for interpretation, without a drive to account for itself. I crave a calm refuge without mind or language.

I dearly love language.

But it often feels like wearing clothes several sizes too small.

Fifty thousand years ago, an elk was struck by lightning and lived. The ache of it stayed in her bones the rest of her life. There was no human there to see it or record it in words, yet it's just as much a part of earth's essential history as any song lingering in a billion human minds.

So why should we think of language as the lifeblood of knowledge or the currency of legitimate existence?

When I think kindly of death, I think of this. I think of rejoining a world that needs neither human words nor permission to be fully embodied in its own existence, an existence governed by ancient, unifying laws of matter and energy, change and movement.

Humanity gets so endlessly lost in thinking that our perception and interpretation of things is the heart of existence, as if an eel or a cloud or the Catskill mountains weren't wholly here and in possession of themselves before we named and categorized them.

But these aren't concepts I would enjoy explaining to loved ones as I tried to justify why they mourned at a stone above an empty grave, why they would never hear my voice again or why I wouldn't be there in the future to help carry groceries in from the car.

And in my most lucid moments I know that it is simple self-deceit to call an elk natural while labeling instinctual human frameworks of understanding as alien and artificial.

The elk and human thought patterns are both children of the same nature.

I couldn't honestly claim that ending my life was an act in service to balance or nature or some philosophical ideal. It would simply be a service to me, a way to escape the suffering of my illness.

The temptation to die was often strong, but my empathy was forever stronger.

Even now I think of death frequently, but a thought is not an action.

As a student, I loved Shakespeare's play *The Tempest*. In the play, when the old wizard Prospero looks to his retirement, he remarks that his "every third thought shall be my grave."

That quote has always resonated with me, and I still have plenty of Prospero days.

Now, however, on hard days, I can look at death kindly and not feel it as a beckoning force, but a gentle, distant rain sweeping toward me from the west, a rain I will find soaking the parched earth at my feet one day. One day. And I can think of that inevitable day without frantically trampling through the fields of flowers ahead of me while rushing to meet those curtains of falling water.

Do I need to tell you that I didn't say any of these things to Dr. Bartman?

"Yes, I'm sure you're busy, but picture an elk fifty thousand years ago ... no, this is relevant, trust me."

Life has taught me that I have an unusual viewpoint, and I have learned to be judicious with how much of it I expose in everyday conversation, especially while under the hum of florescent lights.

So, no, I did not have a plan to kill myself.

I had a set of plans and philosophies for how I would reconcile killing myself with my firm wish not to be a gateway through which more reckless pain enters this world.

I kept my answers short.

"No. I don't have a plan."

Dr. Bartman seemed to check several boxes in her mind that signaled it was time to move on with the conversation.

I now know that the next question, if I had affirmed both my means and my plan, would likely have been, "Do you have a date in mind?"

Those are typically the three key questions for suicidal ideation.

The means. The how. The when.

A "yes" to these three questions, coupled with urgency and specificity, might have called for more direct intervention.

But suicidal thoughts?

A passive death wish?

To a family doctor, these are as common as a runny nose. We just don't discuss the sniffles with the same reticence and shame as wishing we wouldn't wake up in the morning. Our taboo against such subjects robs us of the opportunity to take comfort in our shared struggles and shared humanity. We allow the shame to hush us. We are deterred by the squirrel baffles and the uncomfortable conversations.

Most of us don't have an easy social script at the ready for a friend telling us they want out of this existence.

Dr. Bartman nodded and typed something on my record.

One thing I always appreciated about Dr. Bartman was her approach to discussing medical topics. She simply laid out evidence-based explanations of her recommendations, explained the limits of her own expertise, suggested a referral if needed, possibly recommended a book or article, then listened. She wasn't patronizing. She didn't feel judgmental or dismissive. She acted like a consulting scientist lending her expertise to my goals and wellbeing.

She told me that, typically, the most successful approach to depression like mine involves a twofold tactic including medication and cognitive behavioral therapy, or CBT.

I appreciated the practical simplicity she brought to the subject.

This is a known illness. We have some data about what approaches lead to positive outcomes.

It gave me a kind of abstract hope.

The entire visit and conversation felt hopeful, surreal, and abstract.

My depression doesn't register in my mind like a medical issue. It feels like a natural, organic reaction to a broken world and a flawed personhood. Depression defends itself. So, to sit across from a person in a lab coat telling me about the science behind "fixing" my pain, my outlook, seemed bizarre and slippery. It was like being presented with a simple set of driving instructions for exiting reality as I knew it.

But I knew, of course, that having such a conversation was exactly why I was in that office. I didn't want to be exhausted and in constant physical and emotional pain. Yet it's an odd feeling to speak in a medical context about having a fundamentally new and different life, a new and different outlook. It requires strong suspension of disbelief.

She made it clear that finding a CBT counselor would be my job.

Dr. Bartman simply emphasized that, as she understood the science, CBT was the most demonstrably effective method of treating depression. So, she suggested I narrow my search for a counselor to practitioners of CBT specifically. I dutifully wrote this down and filed it away in a mental drawer labeled "later."

That recommendation made, she turned to prescription antidepressants.

It took an effort of will to hush the old voice telling me that even considering drugs was a surrender, an admission of weakness, that maybe I could revisit the topic once I had gotten myself a bit better on my own. In other words, I could go to the gym as a reward for getting myself in shape.

I swallowed the thoughts and listened to my doctor.

We discussed several types of medication and potential side effects. We settled on fluoxetine, AKA Prozac.

I did not like the connotations the word *Prozac* carried when I rolled it around in my mind. But I was falling from an airplane and dickering over available parachute colors. So I kept quiet and agreed to try the drug.

I left with an errand to pick up my prescription, words of encouragement, and daunting homework to find a counselor.

❖

Forsythia is not a dandelion, and attempting to pluck it from the earth is not a humble goal.

You can cut it back until you have nothing but six inches of knobby wood jutting from the ground, but it isn't dead. Of course not. Forsythia is the masked killer in every slasher movie. It's going to get up again as soon as you look away.

You've probably only made it angry.

Cutting it back will, eventually, kill the plant if you wack-a-mole each returning leaf and shoot in a battle of wills—likely a couple seasons of diligent work.

A little research told me that the surest way to remove a forsythia was to dig out the roots, especially the big, central tap root.

There were seven formidable forsythia bushes in my yard, but I knew better than to overwhelm myself into inaction by thinking of them all at once. I was becoming familiar with that snare on the path to my goals. No, I would focus on just one plant. The most prominent bush, directly across the yard from our big, rear picture window. An explosion of yellow flowers next to the rusting, barbed-wire-topped cemetery fence.

I had two shovels.

I thought of myself as relatively strong.

How hard could it be?

Twenty sweaty minutes into the work, I owned two broken shovels and a growing tolerance for invasive plants.

My wooden-handled shovel splintered and snapped as I tried to pry up the roots using a rock as a fulcrum.

My fiberglass-handled shovel met the same fate.

The flowery whips I pruned away lay in a pile, each severed branch whispering, "Just leave us here on the ground and see what happens."

I thought of the cluster of bare twigs in my grandpa's fist, the twigs that became an entire hillside of tough green branches in a handful of years.

I was exhausted.

I told my dad about the project, and a couple days later he brought me a thick-handled mattock, a tool that is part pick and part axe. He had recently sharpened it on a grinder, leaving gleaming silver edges that stood out against the big, rusted head of the tool. It was rough, heavy, and felt appropriate. A weapon to slay a monster.

Newly armed, I launched another attack that left me winded, soaked with sweat, standing over a jagged crater full of splintered roots and shattered stone.

I sucked wind and plucked the bigger root fragments from the hole before kicking in clods of earth to backfill.

Regarding each root and twig with suspicion, I placed them all on top of a long-dead pile of brush in the corner of my yard, making sure that none of them could reach the soil. It felt a bit like laying a vampire to rest. Buried facing down. A clove of garlic on the tongue. A stake through the heart.

A Carolina wren twittered at me from a neighboring forsythia, bobbing up and down on the branch in the wind, a wag of a golden finger in my direction.

Why was I doing this?

Sure, forsythia is not a native species in Ohio, but was that my project to own in that moment? Was inflexible all-or-nothing thinking really the attitude to serve me best in the tumultuous spring of my healing journey? Was the purity politics of which lifeform deserved to share my little corner of the earth going to be part of my budding sense of connection to nature?

I put the mattock in my shed and sat down on the grass to look around my yard.

Maybe it wasn't an ideal landscape.

Maybe that didn't matter.

Maybe what my yard was, in that hopeful spring afternoon, had more weight and value than what it should be in an ideal world.

I didn't want a yard full of forsythia and blue spruce.

I didn't want to be a person who had a prescription bottle of Prozac on the kitchen counter next to the Mr. Coffee.

But I also didn't want to be someone beholden to all the old, inflexible rules.

The muddy, frantic, awkward explosion of spring growth all around me was chaos. It was also so full of stubborn promise.

Another wren flitted to the shelter of the nearby forsythia.

Nothing at all flitted to the rough patch of splintery soil I had sweated to achieve.

Maybe this isn't a time for ideal, I thought.

Maybe it's a time to learn to love the mess.

I would plant a wild elderberry where the forsythia once stood. That could be enough for now.

I looked at the clusters of gold around my yard and I made an effort to view them in a new way.

As forsythia grows tall, its branches bend beneath their own weight, bowing to the ground in arches of yellow flowers. Wherever they touch the earth, the branches root and send up new shoots, stitching gold across the landscape. Some new kinds of knowledge shift our center of gravity, staggering us, bending us low beneath the burden.

If you think of yourself as a stone tower, this shift is a cataclysm of splintered rock. If you're forsythia, then every startling truth that bends you low becomes a new connection to the earth, a new way to stand, an invitation to grow.

Chapter 7

MORELS

MUSHROOMS ARE THE TEMPORARY REPRODUCTIVE structures of massive, seldom seen, subterranean creatures more closely related to you than to plants. Fungi. Some are miles wide. Some live thousands of years. Some help trees speak to one another. There's magic beneath the forests.

My mom always called the woods "her church."

I'm not sure I understood what she meant when I was small, but I reckon I understand it now.

The woods is a place where we feel grand mysteries and tease out the spiderweb strands that connect us to them and, ultimately, us to ourselves.

It's a place where the sheer beauty of our surroundings kindles a kind of broad gratitude for being awake and aware among such a feast for the senses, such a bounty of incalculable wonders that all share a family resemblance to each other and to us.

Truthfully, I never felt actual churches felt like church, not in the way I knew they were supposed to feel from cultural osmosis, not even when I visited ancient cathedrals in Italy or the famous Abbey where Darwin is buried in London.

No painted ceiling can compare to the actual sky.

No stained glass window can represent what sunlight through new, spring leaves signifies to me, can capture the simple fact of nature's visible existence and invisible meaning.

I suppose, if the woods were our church, we must've had a religion. And we did dutifully observe certain holy seasons, times not tied to the calendar, but to nature itself.

Morel season was one such time.

Long before I had any inclination to eat them, finding morels was special.

Morels, sometimes called sponge mushrooms, are a wild mushroom found throughout the northern hemisphere. They have a conical cap that looks a bit like a sea sponge or a honeycomb, with a hollow interior. There are nearly twenty varieties in North America, though in our inexpert gathering we typically only noted light and dark varieties (we ate both). They have a subtle, nutty flavor.

I knew early in life that morels were a family tradition. My mom hunted them with her mother in Eddyburg. We constantly discussed when they would arrive as March crept toward April.

"It should be soon, but the woods don't look quite right yet."

In this case, the look was mayapples and bumblebees. Early enough in the spring when the fresh green shoots still felt like a novelty after a long, gray winter, the undergrowth alive and moving, but not too overgrown to spot the mushrooms.

I never loved the taste of morels, but I would eat them for tradition's sake and because I liked the idea of physically absorbing both the essence of the woods and the associated mystery of mushrooms themselves.

We would soak them in salt water, slice them in half, dredge them in flour, and fry them in butter. We are Ohioans. If we could pluck starlight from the sky, we would dredge it in flour and fry it butter.

Many folks think of morels as a delicacy, and so some of my joy in hunting them was in delighting others by emerging from the woods carrying a plastic shopping bag full of a treasure that you couldn't buy fresh in any store. As far as I know, morels cannot be cultivated in a reliable, commercial way, so their connection to their natural habitats feels both sacred and essential to their character and value.

Sure, maybe somebody would sell you morels.

But at some point, that transaction involved careful steps through the woods and dedicated searching.

They would need patience.

Perhaps a walking stick to tilt the mayapples aside.

And a close enough relationship with the land to know when it was time to look.

Beyond that, there is all sorts of folk wisdom about morel hunting. My grandmother swore she could smell them, and I've heard others say the same. We had a theory, or perhaps a superstition, that they liked to grow near apple trees.

One thing always seemed certain to me. The ones that grew near my childhood home were unpredictable.

A bountiful patch in one location one year did not mean there would be a repeat performance the next. Except sometimes it did mean that. A reliable acre of woods would be generous one year and then so empty the next that I would doubt my memory.

I can certainly doubt my skill in foraging.

There are expert morel hunters, and neither I nor my family would claim such knowhow.

But the trick of it is that even looking and failing to find anything meant a handful of hours carefully noticing the details of a spring woodland.

No morels, but there is a well-fed toad in that bed of moss.

No morels, but here is a perfect sparrow skull as light as sculpted mist cradled on a leaf.

No morels, but here is a freshly sprouted Jack-in-the-pulpit, and don't I remember an old story about how its dried and powdered leaves could paralyze minnows in a pool?

If the hunt for morels was a holy season in the wilderness church of my childhood memories, then the sacred rites of that festival had little to do with finding and everything to do with searching.

❖

The medication helped.

There were side effects. A suppressed libido. A slight feeling of floaty disconnection like the soundtrack to reality was a hair out of sync.

I didn't spend much time fretting over these side effects. I was too busy feeling grateful and relieved.

It's paradoxical, but the effect I experienced was somehow both subtle and monumental.

I felt a sense of wonder at the enormity of the relief.

It was like aloe on a sunburn.

A sip of milk after a hot pepper.

My medication didn't banish depression from my life, but it dulled a few of its sharper edges. And there was another kind of relief that is hard to describe, a philosophical relief, a new infusion of hope. My pain could be reached through a conventional treatment. I now had evidence that it was possible.

The positive effects were novel data in another way. I would have struggled to see the shape of the pain that had been eased until its sudden alteration left an after image in my mind, a silhouette imprinted on my consciousness. A pain I had no name for had been numbed. Something that registered as both emotional and physical.

My depression was still depression, but the physical desperation and immediacy of it was diminished. The quality of what hurt, and how much, had shifted.

With that new, electric knowledge, a couple promising thoughts began to surface.

If time in nature brought me this far, what else could be done?

If medication eased and defined my pain, what else could be done?

I had been drowning without dying for too, too long, but now, I had flailed in a new direction and my numb fingers had slapped metal.

A buoy floating into my churning, desperate patch of ocean.

The buoy didn't pluck me from the waves or warm my blood, but it did mean that I could think of things beyond drowning. It meant I could raise my head above water and scan the horizon for land. It meant that I could spend my thoughts on things other than barely surviving.

Dr. Bartman had told me that medication in conjunction with CBT was the standard evidence-based approach.

I had softened on the idea of seeking help and had been rewarded.

I had softened on the idea of trying medication and had been rewarded.

Now I was softening on the idea of trying therapy, and I was hopeful, but afraid.

Medication addressed a chemical dysfunction outside my conscious control.

A body issue.

But therapy?

It felt different.

My depression was defending itself, yet again, with its most trusted weapon. Shame. It told me that therapy could be yet another opportunity for embarrassment. Invasive. Intimate. Perhaps even inherently condescending and patronizing.

Oh, you never figured out how to use your own brain? Well, I'll catch you up on what most people learned effortlessly ... but it will cost you.

Moreover, it required my active participation. Swallowing a pill was one thing. Opening my internal life to a stranger was a much tougher prospect.

My meds were starting to provide me with a bit more energy, but it was not energy I was used to having. Certainly I wasn't ready to rely upon it and build plans around it. I looked at my new reserves of fuel with suspicion.

Dr. Bartman had told me that I would likely need to "shop" for a therapist.

Shop suggested trial and error.

Seeking and perhaps not finding.

The last thing a depressed person considering a difficult and painful task wants to hear is that the first effort probably won't be enough. I was facing the difficult process of letting a counselor into my world of sorrow, shame, and struggle and, apparently, I might have to face that process over and over again.

I might search and not find a fit.

I might search and be rewarded only with more searching.

And ... so what?

It was early April and the woods were beginning to look right for morels. An old, half-forgotten holy hour was approaching beneath the trees, along with the bloodroot and spring beauties. I knew I would walk in the woods, and I knew I would scan for morels in the hopes of delighting my mom or surprising Leslie with a new flavor. And if I didn't find any?

So what?

Oh well.

I will have gained more in the looking than I lost through not finding. It was a lesson from another time in my life, a time that was increasingly intersecting with the present.

Why should hunting for a therapist be any different?

I resisted trying therapy for a long time because I thought I was too smart for it.

Here's the thing.

You can't think your way out of depression any more than you can think your way out of drowning.

Asking for a life-jacket is more important than knowing the physics of buoyancy.

Encountering a buoy in the open water is a more vital turning point than philosophizing about burial at sea or drafting not-quite-poetry about uncaring tides.

Expertise is real. And seeking it is as time-honored a human tradition as hunting mushrooms.

After all, what is the use of all these words if we can't share what we have learned in service to one another?

❖

For some time now, I have been publicly open about my mental illness. Doing so has led to many new friends and good conversations. Occasionally, I am contacted by a concerned person who deeply wants to help a loved one suffering from depression.

These can be difficult messages to answer because I sense the desperation behind the words.

Their loved one is withering and they cannot see or, in some cases, understand what is causing it, but they want to do something about it. Anything.

I tell them I am no expert, then carefully offer two pieces of simple advice.

Encourage. And remove obstacles.

You cannot force someone to talk to their doctor or try medication or seek therapy, but you can tell them that you hope they will because you love them.

You can also lend them your energy by clearing a path to these healing interventions.

In the United States, there are many barriers to treatment—although I suspect some of them might not appear as barriers to someone who is not suffering from mental illness.

My advice comes from my experience.

Leslie was my loved one who was desperate to help me.

Her encouragement helped, even when I couldn't communicate that fact to her.

Removing barriers helped.

She would do research and compile lists of therapists in our area that accepted our insurance. She would dig deeper than that and read profiles to try to find good matches for me, excluding people who seemed focused on religion or whose specialties did not fit my struggles. From there, she would send me lists with links and email addresses and contact forms.

She offered to set up appointments.

To drive me.

To wait outside or join me in the session.

In short, through word and deed, she showed me that she was invested in my path toward healing and that while she could not walk it for me, she could ensure that I would not be walking it alone. Her efforts made a difference.

❖

The first counselor I tried was a bust.

She seemed nice.

She was nice.

But she was not right for me.

Part of the problem was that I didn't know what questions to ask. I didn't know that I should ask questions. Nor had I really understood or embraced the idea that I was essentially interviewing this person to hire them for a job. Instead, I had arrived and considered my presence to be my portion of the work mostly concluded. I waded into the gentle white noise and subtle, pleasant scents that filled her office, sat on the comfy couch, and prepared for the healing to happen, for it to wash over me.

We chatted.

It was, if nothing else, practice articulating my struggles aloud. Practice I needed.

I wasn't accustomed to speaking openly about my mental illness and it didn't come naturally. I felt like a fraud. I felt like I was making excuses. I felt my depression defending itself. I kept talking anyway.

I returned to her office three times over three consecutive weeks and with each visit my doubts grew.

Once, she asked if I owned any shirts that weren't black or gray, seeming to imply that my wardrobe choices were a cause of my sadness.

For much of our time, she asked open-ended questions, listened to my answers, asked follow-ups about how I felt concerning the things I said, and moved on without challenging or commenting on what I shared. It felt a bit like beige, office-party talk, except that instead of discussing the weather or college sports, I was talking about being suicidal and sharing things I had only ever discussed with my wife.

The final straw was in the third session when we began discussing my writing.

She was clearly curious about me as a writer.

She shared that she had several book ideas herself.

And then she made the comment that convinced me it was time to move on with my search for a counselor.

"Well, perhaps you need your pain in order to make art."

I didn't stand up and leave the office that moment, but I don't remember anything else she said. I was politely counting the minutes until I would leave and never return.

Firstly, it is an ugly myth that misery is the soul of art.

Secondly, even if I believed such a concept, why the fuck did she think I was sitting in her office?

"Oh, so constantly craving death and making complex deals with myself to summon the energy to brush my teeth is actually perfectly suited to my vocation? Ah. Well. So sorry to have bothered you."

I went home and canceled my future appointments via email.

She didn't ask me why.

She didn't follow-up at all.

I couldn't help feeling a little defeated, but I had learned something.

No morels, but perhaps a beautiful stone to put on my bookshelf.

I now knew what I didn't want.

I didn't want someone to nod at me while I spoke about my struggles and then say, "Your credit card is on file. See you next time."

I wanted to know, up front, how my counselor would approach my specific issues and I wanted to be let in on the logic, history, and science behind their method.

I wanted a team member, not a sympathetic ear.

I wanted an expert, a tutor, a mentor, not someone who would simply tell me that my feelings were valid and offer clichés or toxic positivity as the entire substance of their contribution.

I had learned a valuable lesson.

So, every time the belittling voice in my head would whisper, *back to square one*, I could shake it off and reply, *no, not exactly*.

No mushrooms. But something.

❖

There were a handful of other snags on my path to finding a counselor, but no more fruitless appointments.

One person I contacted was retiring.

One was booked for the next seven months.

One no longer accepted my insurance.

Finally, I contacted a place on Leslie's "accepts our insurance" list called Northwood Clinic, about twenty minutes south of my home. The arboreal name seemed auspicious at least.

I used the general contact form, which put my request and a brief summary of my struggles through to the psychiatrist who runs the practice. He passed me on to a counselor named Mark Travis, and we scheduled an appointment.

I did not want to go, but I went.

I was developing a nascent confidence in my ability to distrust my own distrust.

I arrived.

Another tidy brick building.

Another waiting area with white noise machines and soft, gentle everything.

I sat and stared at my phone, twenty minutes early for my appointment (which is my version of on time), and wondered about who I would be meeting. I had seen a tiny picture of Mark on the Northwoods website, a cheerful-looking bearded man in his sixties, but when he walked through the door he wasn't what I expected.

Mark looks a bit like he should be playing in a band at a biker bar (a thing I know he has done) or standing by the counter at an army surplus store discussing improvised survival shelters (a thing I suspect he has done).

Army green jacket.

Jeans.

Boots.

Long gray hair and a salt-and-pepper beard.

He greeted me, then gave me the "one moment" sign while he disappeared deeper into the office to complete some business.

He returned. Unlocked his room. Slipped inside and changed from his army jacket to a more formal-looking vest in a now-familiar ritual that I think of as shifting to business-mode after he has enjoyed a cigar in his truck on the drive to Northwoods. Mark sees a handful of clients in the evening after working his primary job at a semi-residential treatment center, where he presides over a variety of outpatient group therapy sessions.

Newly be-vested, he welcomed me in, shook my hand, and gestured to the exact neutral brown couch I expected to see.

I took a seat.

He held a notebook and a stack of paperwork in his lap.

Mark smiled and spread his hands.

"What can I do for you?"

It was a big question.

"I am ... struggling with depression. My doctor recommended a two-fold approach. Medication and CBT counseling."

Mark nodded. I continued.

"Could you tell me a bit about how you approach CBT? About your counseling practice? And, well, maybe define CBT for me?"

These were the things I wished I had asked my last counselor.

The questions seemed to please him.

I had introduced topics he was planning to discuss as part of his first session routine.

"Yes. CBT is how I work. I can describe the process and how I approach it, then we can decide if we're a good fit to work together."

I liked his tone already.

My first counselor never offered me any explanations of her process and never suggested that we were determining compatibility.

Mark was interested in my informed buy-in to his process, in my active participation and collaboration in his approach.

He explained that CBT essentially works on the deceptively simple premise that our thoughts shape our emotions, which in turn informs our actions. On its surface, that connection between thoughts and emotions seems obvious and basic, but as soon as we started unpacking the concept, I felt I disagreed.

In my experience, it seemed emotion came first and that my thoughts were how I discovered the source of the emotion. I was sad, so I could use my intellect, my thoughts, to find the source of the sadness, to solve the mystery of my seemingly sourceless despair. The usual suspects for the cause of my sadness were my flawed self and an objectively sad world.

Mark challenged my assumption, arguing instead that the sadness was springing from a thought. He argued that emotions are not senses. Feeling sad or angry or hopeless is not revealing anything about objective reality. These feelings are not like a scent, sight, or sound.

"What emotions do is interpret our thoughts."

I had difficulty with this premise.

It was particularly difficult because I didn't seem to be having any thoughts at all as I felt the tides of pain and sadness ebbing and flowing inside my skull, constantly eroding my hope and energy.

My thoughts, when I noticed them at all, felt reactive.

They didn't feel like the cause of anything.

I wish I wasn't miserable all the time. Why am I so broken? What could be redeemable in such a flawed, hateful world? Amidst a climate crisis of this magnitude, what it the use of seeking happiness anyway? Is it even just or reasonable to be happy?

I believed such thoughts were interpretations of my sadness, not the cause.

CBT disagreed.

Mark pointed out that there are terminal cancer patients and folks struggling on the edge of starvation or fleeing from war who call themselves more content than me, a relatively healthy, middle-class man living with innumerable privileges and support, a man who frequently wished for death as an escape from his suffering.

There is no objective constant for happiness. No infallible, measurable predicter for subjective states like contentment or hopelessness.

Different people can and do have widely differing emotional responses to the same events and conditions.

How could that be true if emotions arise from the bare facts of our circumstances? If they are born of our lot and context and not our subjective interpretations or thoughts?

"Well, perhaps I just pay more attention than other people?"

"Oh? So happiness is born of ignorance, and sadness of awareness?"

I knew that wasn't correct. I knew the moments of happiness I had experienced in life did not exist because I had forgotten about loss or suffering or injustice or the truth that all things are impermanent. I also, at my core, did not actually believe that despair was truth's native, natural emotion, even in a flawed, temporary world.

It wasn't true, and I knew it wasn't. I also knew that I had often embraced such logic as a reason not to seek help, accepting hopelessness and shaping my worldview around it, changing my understanding of reality so I wouldn't need to change anything about my own inner life.

Except I was currently seeking help.

I sat in that office in order to actively invite change to the party.

And in that context, in that room, it was clear how transparently incorrect my line of thinking had been. No, I wasn't unhappy because reality is unhappy. Unhappiness is not wisdom.

Everything ends. All life. All worlds. All species. All achievements and efforts and carefully crafted identities. Nations. Religions. Philosophies. Languages. Civilizations. There is no pleasure, peace, or contentment that can ever represent an achieved final, static state of fulfillment, of happiness captured once and for all.

And?

So what?

Should I not enjoy a meal because I know I will eventually be hungry once again?

If I write a poem that touches me and connects with others, was that project of no value because the language I write in will not survive as long as the sun under which I wrote the words? Does the pleasure from that poem need to be "correct" in an objective sense? Does it need to have objective merit on a cosmic scale?

Says who?

Happiness itself is a kind of meaning, to be chosen and crafted much as I chose to collaborate with the heron on that pale winter day when I realized that natural wonder had not disappeared from the world simply because I had stopped seeking it.

When I cried seeing that wading heron, I wasn't crying because of the heron alone. I was crying because of the beauty in what the heron and I did together, because of the relief our meeting brought into my unique, subjective world that exists for and within and *because of* me.

Emotions arise from our subjective interpretation of our circumstances and context—in other words, from our thoughts.

I had been wrong.

I began to understand the distinction, but it remained uncomfortable to accept.

The idea that my thoughts were manifesting my emotional pain felt accusatory, like I was being told that I invited my own suffering, like the fundamental problem was that I just kept touching a hot stove and couldn't solve the mystery of why my fingers hurt.

I said as much to Mark and he challenged me again with a simple question.

"Are you choosing to be sad?"

"No."

"There you go."

It was confusing.

He sat there in his vest, a notebook on his lap, tugging at the threads of my well-worn thought patterns. I couldn't decide if I wanted to hug him or break his nose.

He studied my expression, shook his head, and found another angle of approach.

"Part of learning to use CBT is learning the skill of noticing our own thoughts. In order to shape our thoughts, and understand how they interact with our emotions, we first have to catch hold of our cognitions and see them clearly. And I call it a skill because it's learned and gets easier with practice. It's neither easy nor intuitive. Not when we first begin. It's fucking hard."

As I began to understand, I saw the utility of CBT coming into focus.

If I could become aware and intentional about my own thoughts and thinking patterns, I could, in theory, begin to actively influence my emotions and actions.

During my visits to the woods to hunt for morels, I went armed with the knowledge that, more often than not, I would find nothing. Why didn't that thought stop me from going in the first place or make the experience a sullen trudge? Perhaps because I could hear that thought clearly and refute it. I had practice sensing and refuting it. I could talk back to that bitter line of thinking and say, "I will choose to enjoy what I find, even if I don't find the thing I seek."

I knew intentional thought could influence my emotional outlook. I had experienced it when I hushed my anxiety about potential disappointment when mushroom hunting. So, was it really so farfetched to imagine that there were bitter, self-defeating thoughts I couldn't sense and couldn't answer lurking in my mind? So much of what a mushroom is happens down in the dark, beneath the leaf litter.

Mark continued.

"Our minds have all sorts of clever ways to mask unhelpful thoughts as plain reason. Thought distortions. Self-deceiving falsehoods. Things like being certain we can read other people's minds after we leave a party or sense inevitable doom waiting to meet us in our futures. Things like all-or-nothing thinking or filtering positives out of our awareness. But we can't interrogate our thoughts or name their deceptions until we learn to recognize and acknowledge them. Until we perceive them clearly."

I could see the logic in it, but I felt skeptical, and something about Mark's simple, candid way of speaking made me say so.

"And if I can't? If I can't learn to do it?"

He gave me a steady look.

"Then we can't work together."

I stared at him.

He meant it.

I could feel that he meant it.

He wouldn't lose sleep if we couldn't work together. He'd been doing this for forty years. But he also wasn't being cruel or meanspirited. He was being honest and straightforward. He was saying, "Here is a tool I can teach you to use, but if your starting position is certainty that you can't learn it, there isn't much more for us to say on the subject, is there?"

"Okay," I said. "I think I can do it."

He shook his head.

"No. You will do it."

"I will do it."

I could say those words. But they didn't feel real. Not yet.

I tried to push back against my own rejection of Mark's description of CBT, and I felt my illness sinking its claws in to anchor itself in place. It hurt. So, I retreated to a safer thought that still felt productive.

Depression defends itself.

Depression defends itself.

Depression defends itself.

I felt tears coming and I stopped them.

I had many years of practice not allowing myself to cry in front of anyone, especially not in front of other men.

Mark noticed.

"You have a pretty strong inner critic don't you?"

I chuckled and rubbed my eyes.

"Yes, I think that is fair to say."

"What about your inner cheerleader? Do you have one of those?"

I thought about it and couldn't conjure up an image of such a person. Mark nodded.

"That's something we can work on too."

I swallowed. My throat felt tight. I moved to change the subject.

"I'm taking medication. I think I can feel it softening the pain."

"Good."

"Where does that fit in to the CBT equation?"

"Well, there are certainly physiological factors that may be involved in depression, and medications help a great many people. The way I look at it, in terms of biology, the meds can get you back to neutral. An even footing. The question is, 'Then what?' CBT helps develop the skills to answer that essential question, especially when you feel you have lost the skill of happiness and contentment. Finding that neutral footing is often just the beginning."

Mark's words felt true. And they made me exhausted.

It had taken so much courage and effort to get myself back to the woods, to the doctor, to the couch in Mark's office, and now it was time to think of it as the beginning of a journey, the commencement of another search.

Why couldn't they just figure out how to cultivate and farm morels?

Why couldn't I just buy them at a store or subscribe to a service that delivered them to my loved ones once a month?

Why wasn't there a pill that would make meaning for me, that would circumvent the need for approaching contentment as a skill?

"What do you do for fun?" I asked.

Mark smirked, perhaps recognizing the question as plea to take a break from the hard stuff.

"I camp. I hike. I like to sit by the fire. I make my own maple syrup. I get to the mountains a handful of times a year. I hunt, but I'm rarely successful at it. But it doesn't matter because the important part is the feeling I get just being out before dawn, getting to feel the shift from night to morning happen under the trees. I read. I recently made my own blacksmithing forge."

I smiled, swallowed, and looked out the window.

Yeah. This counselor felt like a better fit.

Storebought morels wouldn't feel like morels.

You can't buy them in a store, because that isn't what they are.

They aren't just mushrooms.

They are the search.

Not possession. Discovery.

They are how we welcome back the green beneath the trees.

They are dynamic. An action. Effort and intention. A conversation.

Morels are everything else you find as you tilt aside the mayapples and willfully choose to believe that next patch of sunlight between the black walnut trees looks promising.

We don't always search in order to have.

Sometimes, we search in order to find.

And there is a world of difference between the two.

This embrace of the process, of the nourishing virtue of the search itself, is a hard thing to accept when what I'm searching for is an escape from pain. I cannot step back from myself far enough to accept that "the journey is the point," not when the journey is driven by so much agony. I can forgive myself for this. Yet, it is vital to understand that the search and the journey are messy, nonlinear things, and that, by virtue of all that capacious uncertainty, they may hold unexpected benefits and opportunities, things I would miss if I dismissed the whole process as hateful.

I didn't know that I was searching for Mark.

He wasn't the sort of person I could have anticipated.

Many wonderful things are beyond our ability to predict.

The world is wide, wild, and full of surprises.

Hunting for mental health is often a lot harder than hunting for morels, but we must honor the world's potential by refusing to call our search a failure before it begins. We cannot know that we will fail. Often, we cannot even know what it is we hope to find.

Chapter 8

WHITE TRILLIUM

I THINK OF THE WHITE TRILLIUM as a valley flower.

I tend to encounter them on tree-shaded slopes deep in the woods.

The species I know best are also called the large-flowered trillium, but I have always loved some of the folksier names, including the "wake-robin" for their early spring blooms and the "wood lily" for their smooth, white petals. I have heard them called "snow trillium," but that name mistakes the white trillium (*T. grandiflorum*) with another flower more accurately known as the snow trillium (*T. nivale*), which is one of Ohio's first bloomers and may be found even among a literal coating of March snow.

The white trillium is a triumvirate lifeform, their leaves, petals, and sepals all clustered in threes.

They are ghostly, fleeting things with milk-white, pointed petals enfolding a golden center.

Ants distribute their seeds, and white-tailed deer consider them a delicacy.

In many ways, they are the antithesis of forsythia.

They are shy, native, deep-woods dwellers who called these valleys home long before humans or gardens or lawns arrived on this continent.

Ohio's state flower, the white trillium is of this place, an old inhabitant of the woodlands where I was raised and still call home. Knowing this brings special awe to seeing a hillside dappled with their sharp, bright blooms. I feel the added depth of time weighing on the scene, tugging my thoughts back through the millennia, pulled by the dizzying awareness that I likely could have wandered upon the same gathering of woodland phantoms, of ghostly flowers, five hundred or fifty thousand years ago.

I find them startling.

Unlikely and mysterious.

Elegant and graceful, they stand in solemn, silent ranks and then are gone before the summer is fully underway, before the year is truly awake.

Morning prayers. Vespers.

White trillium may be ancient and common, but they are also fragile. They don't have long to spend in the sunlight, gathering the energy they need for next spring's bloom, storing it below ground. So, picking one may end the plant's life, starving the root and ending the cycle of growth and dormancy, feasting and sleep.

I think I knew this before anyone told me.

Trilliums look like an offering from the fairies, not quite real and a bit too good to be true.

Any lover of folklore will tell you that offerings from fairies have consequences.

Any worldview that says, "I found it, so it's mine to take," also has consequences.

Still, the trilliums persist.

Strong and fragile. Fleeting and ancient. Temporary and yet, at the moment of this writing, very much here.

So much apparent paradox in nature is not paradox at all. At least, the paradox is not rooted in nature. It is rooted in us, in our narrow understanding of strength.

———

The trillium is older than humans, yet it needs the ants.

It needs its fleeting season in the sun.

It needs favorable happenstance to deliver it from hungry deer and plucking fingers.

Many seeming paradoxes in nature seem so because they feel discordant with our cultural sensibilities. When this occurs, it is not nature that is out of step with reality. We are being offered a lesson, if we are wise enough to accept it.

❖

The day after my mom's cousin Cathy committed suicide, a thank-you card arrived in my parents' mailbox.

The card was from Cathy, sent the day of her death, thanking my mother for her many kindnesses.

One of her final stated wishes was to have her ashes scattered among her favorite flowers, the trilliums.

A wish that, last I heard, had not been carried out.

I didn't know her.

I primarily know the way her passing touched my mom.

Yet, some aspects of her story make me prickle with recognition in a way I find useful to explore.

Mom told me about the thank-you card on a day I stopped in for a visit while Leslie was at work.

I say "stopped in." They live an hour away, but I enjoy both long drives and my parents' company.

Mom said the card felt "creepy."

Understandable.

I said I thought it seemed sweet.

Our conversation on that day felt like strange destiny. Despite my luck with medication and my recent success in finding a counselor, I had been struggling more than usual with intrusive thoughts about suicide.

Later, I would learn that my struggle was not unusual for someone newly on medication, as the threat of suicide has a seemingly contradictory tendency to increase after early successes with treatment. It's the unfortunate consequence of gaining more energy and clarity before gaining the skills to challenge the hopelessness of a depression-shaped inner life.

Even so, suicide wasn't a subject I discussed openly. Certainly not with my parents. And here it was, the topic that felt like a dark and secret shame stuffed down deep in my pocket, now sitting on the kitchen counter in the sunlight next to my coffee cup.

"We all knew it was probably inevitable," Mom said.

"Was it? Why was that?"

"She hinted about it. And she suffered from several kinds of cancer, along with other health problems."

"And the treatments weren't helping?"

"As far as I know, she didn't really seek treatments. She wouldn't. Said it was too expensive, among other reasons."

"Why not just get the treatments and then not pay? Or arrange payment over time?"

Mom shook her head. "That wasn't how she thought about it. But I don't think it was just the cost. She also wouldn't let anyone into her house, and when they finally went inside after her death they discovered a hole in her roof."

"A hole?"

"Yep. Letting in rainwater. Mildew and mold growing. The floor was rotting beneath."

"Well. I suppose those were her choices to make."

Mom nodded and looked sad.

"Yes. Although I know she often ignored things and then demanded help from her brother. I think she could be a bit difficult to love."

"I can imagine."

I thought I could imagine.

I also felt difficult to love.

I thought of the days when I lay face-down on my floor and glimpsed Leslie's expression when she saw the state of me. I thought of the months stretching on with me as a morose, live-in burden to the woman I loved. I thought of the dead weight of my illness scraping along the ground behind my stubbornly trudging family as they dutifully dragged me into the future, while I, unable to carry my own bulk, unable to participate in my own healing, couldn't bear to face the steps needed to improve my own lot.

Mark, my new counselor, was already becoming a voice inside my head. He seemed to perch on a little stool somewhere just behind my eyes. I saw him raise a brow at these uncharitable thoughts about the burden of loving me.

"Is that true?" he asked.

He shifted and his stool creaked.

"No," I answered. "It's not."

It wasn't true.

Not entirely.

I had already accepted help and participated in my own healing.

My fears and doubts were not my actions. My actions were my actions.

I had allowed Leslie into my inner life.

Which was frightening.

I had invited nature back into my worldview.

Which was painful.

I had gone for walks, consulted with my past selves, dusted off old recipes for meaning-making, visited my doctor, risked shame and disappointment by trying medication, and slogged through the frustrating process of finding Mark and learning to say my true thoughts aloud in his office.

Every single one of those incremental steps was difficult, uncomfortable, and often terrifying.

But why?

Why were all of the steps I had taken so hard?

I didn't know Cathy, but I liked her as soon as I heard about her wish to be laid to rest alongside the trilliums at Blackhand Gorge.

I liked her and I found I could easily imagine reasons she wouldn't ask for help.

I had many advantages she did not, and I struggled daily to ask for and accept help.

I could imagine what she thought it would cost her, whether or not she could articulate such costs.

It is already difficult and shameful to feel like a burden.

That feeling makes asking others to share in the responsibility of that burden, be it emotionally, physically, or financially, all the more repugnant.

"How could I inflict this, inflict me, on others?"

Seeking and accepting help is a skill, and it's a skill that we are not taught to value. In fact, we are taught to shun it.

In the culture that raised me, rugged individualism is praised. Summoning the courage to admit vulnerability and seek aid is an admission of weakness. Rugged individualists do not need help. They win or die.

We are the culture that took the phrase "pull yourself up by the bootstraps" and twisted it to mean the opposite of its original intent. Originally, that saying was meant as a joke, as an example of something as ludicrously impossible as pulling yourself over a fence by your own bootstraps. Now, we forget the origin of the old saying and use it to suggest that we can succeed all on our own, an impossibility that we have decided to enshrine as a cultural value, ignoring the implicit absurdity of both the metaphorical act and the idea that we do anything alone as members of a society or an ecosystem.

I am comfortable with death.

I think about it frequently and have made peace with it in a variety of ways.

Still, to me, suicide remains a tragedy.

It isn't a tragedy because a life has ended. All lives end.

It is a tragedy because every example of suicide that has touched me personally seems to have one key thing in common. The person chose death because they felt trapped in an intolerable situation that they didn't know how to escape. They didn't have the tools or ability to ask for help. They didn't learn the skill of productive, honest vulnerability. Support was absent, out of reach, or beyond their ability to seek.

When we are not taught that asking for help is a courageous act, all of our weaknesses become shameful liabilities. Being needful, of anything, becomes a sin, a character flaw. Perhaps we all know, on an intellectual level, that no one succeeds without prodigious help and, consequently, no one is truly independent, yet, the logic of that obvious truth does not lessen the emotional weight of a cultural belief system that links seeking assistance with shame.

Many of us would endure absurd levels of pain and personal risk to avoid shame. Suicide is a danger to our bodies. Shame is a danger to our identities. It is not uncommon for people to sacrifice their bodies when their identities feel threatened or damaged.

I had been earnestly considering suicide for years. Was I unaware that there were available treatments for depression? Could I honestly say I was certain that none of the treatments could help me?

No, of course I couldn't know if counseling or medications would be effective.

What I could absolutely know was that, within my own belief system, within the morality I had internalized, death seemed preferable to admitting I could not overcome mental illness on my own, to admitting I could not manage my own house and affairs. I could know that oblivion was a better alternative to owning the identity of a weakling or a lesser mind. I could be certain of that because the certainty was philosophical, meaning in this case that my certainty provided its own evidence. After all, in the end, we get to decide what our lives mean.

What sort of world taught me this version of morality?

We understand our world and ourselves through stories. Villains and heroes. Successes and failures.

Where are our stories teaching us that it is an act of courage to seek help?

Where are our stories teaching us that coming to another for aid is a gift to them as well as to ourselves?

Where are our stories celebrating the fundamental truth that the trust involved in open vulnerability is the lifeblood of any healthy community and the foundation of vital mutual aid?

I can't recall such stories.

I can recall stories aimed to teach me that debt is immoral.

I can recall stories celebrating the virtues of minding my own business.

I can recall one-man armies and lone survivors. The independently wealthy. The self-made man.

Commonplace fantasies are lessons in shared cultural aspirations and values. The problem is that these aspirations of independent power are antithetical to both the physical reality of our interconnected planet and to the character of the human animal.

Humans are social creatures.

That is our nature and our real strength.

Foals can trot shortly after birth. What defining action can a human do shortly after birth? Only one. Cry for help. We do not have claws or fangs or wings. We have cooperation. We have the ability to seek and give aid.

When we admit we need other people and other creatures in order to be safe, we admit that we are not fully in charge of our destinies. We admit that we live in a world of the unknown and the unknowable. We are forced to come to terms with the fact that we are not gods or masters, but rather fleeting lifeforms and fellow passengers bound together in the same ship.

Acknowledging this reality and learning to live in it are two separate things.

Asking for help is a skill that is both hard to learn and much maligned. It is also a matter of life-or-death importance, as well as a doorway to larger truths about our relationship with nature and each other.

❖

My mom's cousin couldn't countenance debt. She couldn't tolerate certain kinds of help or medical intervention. I suspect that, as she aged, the risk/reward math of making any fundamental life changes became more and more daunting. Perhaps she mailed that thank-you card because she also couldn't stand the emotional vulnerability of thanking my mom in person while still sharing the world with her. None of these ideas feel alien to me.

As Mom described what she knew of Cathy, I could easily imagine finding myself in a similar situation, isolated and out of alternatives, certain that death was the only exit left to me and grateful to have even that option. In the context of my depression, it has been sorely tempting for me, on many occasions, to turn away from the obligations of love and human connection, to turn inward and reject the outside world of constant risk and uncertainty.

Is it so ridiculous to choose death over swimming alone against the hostile currents of our world? Truthfully, I suspect that many of us can imagine ourselves coming to that same conclusion. I suspect that's why the subject makes people uneasy. In modern America, many of us live in worlds of shut doors and private agonies. Choosing escape does not seem so unimaginable in such a world.

And therein lies the tragedy.

Among the half dozen or so suicides and suicide attempts that have intersected my path, nearly all of them feel connected to the unnecessary silence and isolation that plagues my culture.

Depression defends itself.

Cultural values that prevent seeking help hand depression so many weapons.

Cultural values that teach men neither to ask for aid nor to express their emotions hand depression so many weapons.

Cultural values that teach women that they should be silent, ornamental, tireless engines of emotional labor hand depression so many weapons.

Cultural values that tell us the living world around us is an inert thing to be consumed or disregarded hand depression so many weapons.

Cultural values that equate wealth with worth hand depression so many weapons.

And, as depression stands strong atop its arsenal, the bills pile up. Health issues we can't afford, or refuse to acknowledge, worsen. Inconvenient family connections fade with distance and distraction, or actively do us harm through abuse. Adult friendships, strained by busy schedules, become burdensome and fizzle. We work and we buy and we try to build our lives into the little walled cities we are sold as ideals. We look to our mythic figures, to the political strongman, to the self-made billionaire, to the effortless glow of the social media influencer.

"Self-sufficient."

"Self-supporting."

"Self-contained."

We make these falsehoods our truths.

We learn how to be alone. We learn how to shut up. We learn how to be numb. We learn how to have conversations about nothing. We learn how to keep to ourselves, even among company. We learn to pretend we take care of ourselves all on our own. We learn to see other people as competition. We learn to see relationships as transactional. We learn that to be human is to be, above all, separate and distinct from nature, so that even the planet on which we live feels like one more indifferent stranger glimpsed across a parking lot.

We pluck the trilliums, and we do not stop to wonder if they will return next spring. Our culture doesn't cherish the next spring. It cherishes our ability to pluck with impunity, to own without knowledge or responsibility. It tells us that's all there is.

Then, one day, we find ourselves utterly alone within ourselves. A place made entirely of locked doors we can't recall how to open. A place of muted voices from the next room, voices we can't quite make out. A place where the old myths we were sold as blueprints for our identity are getting harder and harder to accept as they clang against our fundamental natures. A place where only one final exit seems available.

Is it surprising that, in 2020, the CDC reported suicide ranked in the top nine causes of death for Americans? Is it surprising that a suicide death occurs about every eleven minutes in the United States? That in 2020, 12.2 million Americans reported seriously considering suicide?

No. I don't think it's surprising at all.

Humans need help, yet too many of us have accepted the idea that it's a personal failing to seek it. We are social animals instructed to loathe our own greatest strengths, communication and cooperation. We are born of the Earth, of iron and water and oxygen from trees and phytoplankton, yet we impoverish ourselves with the idea that only human life is worthy of honor and love.

We internalize these corrosive values and stitch them into our identities.

It is amazing what we will tolerate in order to avoid leaving "Who am I?" lingering as an open question. If someone offers us a premade identity, it is hard to resist taking it.

Making meaning is work, and many, many people, systems, and institutions fight tirelessly to sell us the idea that the only real meaning is the kind they make for us. Name-brand meaning. Storebought, not homemade.

So, nature becomes just a thing.

We strap on plastic Superman capes, call it reality, and step to the edge of the precipice.

The idea of stubborn personal independence feeds our egos at the price of our spirits, our wellbeing, and too often, our lives.

I don't believe that mainstream American culture invented pride, but I certainly think we have honed it to a razor's edge and failed to recognize that it cuts us at least as often as it cuts in our service. Yes, we produce most of the world's billionaires, but we also train people to choose to die alone rather than accept mountainous debt for medical treatment. Our technologies keep advancing, but do we translate those advancements into better, kinder lives? No.

When you build a culture dedicated to the worship of impossible things like unassailable strength and inexhaustible wealth, there is no such thing as enough. There is no such thing as the proper time for rest or healing or art or connection. There will only ever be the chase for unchecked growth and acquisition. We cannot worship hunger and then hope for satiation.

I strive to avoid spending too much time on such generalizations and broad conclusions.

Broad conclusions lack the truth of specificity and, more importantly in the context of my mental health, I cannot act upon grand observations and big societal woes.

I don't control culture.

I'm not in a position to steer society.

I cannot edit our broken definition of strength in the popular imagination.

I'm limited in my scope and influence.

Accepting my limitations is part of my healing process.

Even so, I think such broad observations are important to understanding why Cathy couldn't stand debt, or even asking for help when the rain began pouring through her roof. The big picture is important when I try to understand why my own death seemed clearly preferable to a doctor's visit or a phone call to a mental health clinic.

Suicide feels like an escape when all else is broken, our lives, ourselves, our connections, our stories. Death feels like a way to gain a final measure of control in a world without options, without help. The problem, of course, is that the lack of options tends to be an illusion. Help is available, and seeking it is neither weak nor immoral. It is natural.

❖

Trilliums belong to an evocative category of plants sometimes called spring ephemerals.

After an ant accepts the nutrient-rich food bribe attached to the trillium's seed, a fleshy structure called an elaiosome, it carries it back to its nest. The food is eaten. The seed is abandoned beneath the earth, well-planted and safe from predators. Germination takes two years. The young plant takes between seven and ten years to reach flowering size. After blooming, the plant will live another decade or so, if it's not picked by a hiker or eaten by a deer.

The trilliums can't command the ants.

They can't declare war on wandering deer.

They do not strategize about where to grow.

Happenstance must be very kind to achieve germination, and the decade following must also be kind if the plant is to reproduce.

When we talk of individual strength, when we make it a chief ideal and stigmatize seeking aid, we are falling prey to myth. It's not that individual strength isn't a real and valuable asset. Individual toughness, cleverness, grit, creativity, and willpower are all real and useful things. Yet, they are small things and far from the heart of human power.

We love to pretend that individual power is the key unit of power, perhaps because individual ownership is our key unit of possession and achievement in modern America. We need individual power to be of great consequence in order to justify the way we are taught to think of ownership, wealth, and property. If individual strength is not the key, if communal strength is at the root of all real achievement, then the injustice of our profound financial and political inequality seems ridiculous. Which, of course, it is. So, a great many institutions of inequality and the status quo work to prop up the idea that individual power is our most vital and honorable mode of natural strength.

But the trillium were here first. Their essential strength isn't a matter of individual plants with unique, isolated virtues. Spring ephemerals aren't an accident of nature, and it's not blind luck that accounts for their ancient presence in these woodlands.

Humans don't stand outside the interdependence of the natural world, not in terms of our physical health and not in terms of our societal progress and wellbeing.

When we privilege individual strength and wealth over communal wellbeing, we rob from our natures to give to our egos.

The story of my mental illness is not a story about societal shortcomings, but those shortcomings are an integral part of the setting.

Would Cathy have died in a society in which healthcare is a right not a privilege?

Would I have constantly looked upon suicide as a friend in a society that encouraged open, natural collaboration and didn't depend on isolation and mythologies of individual exceptionalism for its economic underpinnings?

Would death by one's own hand be the ninth leading killer in a society that didn't run on artificial scarcity and willful, manufactured desperation?

I have heard some say that this is the way things must be or we wouldn't be motivated to do anything. Where is the motivation if I might not end up owning more than my peers? Where is the motivation if the penalty for being too poor wasn't dying on the street?

And yet, excessive wealth and conspicuous inequality as a social value is rather new to the history of the human animal for anyone to suggest that it's a fundamental feature of our survival. How can anything as new as barely regulated capitalism be considered fundamental for humanity?

If you struggle with emotional pain and are reading this and thinking, "But I have no power to reshape modern Western culture," then perhaps these pages feel daunting.

I understand.

And, you are correct.

But though our cultural context is part of our story, in terms of suicide and mental illness, it's not the essential part. We don't control it, but there is so very much we do control.

Neither our motivations for living nor our suicides are put to a public vote. They don't result from market demand or the recommendations of PR firms. We aren't required to sell our reasons for staying alive to anyone but ourselves. We are the gatekeepers of our own meaning.

This is good news.

It means we can recognize the ways in which our culture's priorities are broken and decide, for ourselves, to develop different priorities.

It means we can derive pleasure and purpose from seeing a trillium, rather than leasing our purpose from a corporation or a political affiliation.

It means we can realign our strength with the strength of the natural world, a strength that says when we share our risks and burdens, we all carry a lighter load.

Perhaps I am too ignorant on the subject to call myself anti-capitalist, though I am unequivocally against some of the cultural values that seem tied to capitalism. It's the only economic system I've experienced, so I may not be in a position to make honest comparisons. From antibiotics to the airplane, many good things have sprung from the soils of capitalist nations, "from" though not necessarily "because of." Perhaps there is a version of capitalism that actively avoids making greed a chief virtue, that sets robust moral guardrails on competition and demands a better standard of living for all members of society by simple virtue of our shared humanity and a nuanced understanding of the communal nature of any civilization's endeavors. From traffic laws to food safety regulations, a majority of us understand that setting aside some personal freedoms and resources in the name of our communal good is a worthwhile tradeoff. The debate lies in the degree and the details.

I know there are many staunch defenders of American capitalism and they have real examples of beneficial outcomes to support their position. Still, I don't think it's a radical notion to say we can do better. America may be the wealthiest nation, but a quick look at our global rankings in education or heath care outcomes leads me to think it's time to adjust both our ambitions and our understanding of what constitutes real national "wealth."

The survival of any species or society is mutual.

The survival of any ecosystem is also founded on interdependence.

Community is vital for our wellness, but community only functions when we embrace the vulnerability of honest sharing and open communication. This means providing help, but it also means reframing our need to seek aid as a communal good.

Many of us would offer to shovel snow from our neighbor's driveway without a second thought, but how many of us would ask that same neighbor for help shoveling if we sprained our ankles? Providing help may be a cultural value (though I fear even this is imperiled of late), but asking for help remains a firm taboo.

We can reject this.

We must reject this.

Our glorified isolation and manufactured desperation are both deadly and unnatural.

A trillium needs the ants. It needs the forbearance of deer and hikers. It needs the shade of the trees. It needs the sunlight and natural systems that enrich the air and soil. It needs the generations that came before.

Consider the needs of a human child and see clearly that we humans need significantly more than this fragile, ancient, spring ephemeral.

Perhaps it's not just happenstance that accounts for both us and trilliums sharing this world.

Perhaps there is more artistry and warmth in the methods of the universe than that.

I think so.

I think, intentional or not, there is kindness woven into the fabric of nature.

There is certainly kindness woven into every human life; it's a key component. Our individual survival demands it as surely as the trillium needs its woodland community.

Caring for a child is profoundly hard work. There are over eight billion humans on Earth as I write this. Imagine the magnitude of care, patience, and kindness prerequisite for that number.

I am sorry that Cathy chose to die.

I am sorry to have lost friends to suicide.

I am sorry for acquaintances and loved ones who have attempted suicide, and I am sorry for everyone who has found themselves looking on death as a friend and last, best option.

Everyone is different and so our needs are different, but I feel confident these people did not need "me versus the world" myths of individual toughness in the service of individual glory.

They needed the skill of asking for help and a culture that wouldn't demonize their asking.

They needed support, and they needed options.

They needed role models of honorable vulnerability and new archetypes for making meaning.

They needed stories in which a wounded mind or an atypical selfhood is not a broken kind of being or a two-dimensional, villainous identity.

They needed the simple, traditional strengths of the human community and the lessons of our interconnected planet.

Worshiping individual power and wealth betrays the truth of what we are and risks all that we could become.

In the end, strong or weak, rich or poor, we are all spring ephemerals.

Too many people look at a maple tree and see only the hard wood, not the soft, fluttering leaves that built all that strength and rigidity.

Trillions of leaves, tender and green, live, die, and fall each year. So fragile. So temporary. So essential to every towering tree that bathes in the starlight of centuries. It is an old truth of nature that softness often holds the key to enduring strength.

If needing aid is soft, it is the softness of a leaf.

Chapter 9

EASTERN BLUEBIRD

RETURNING TO THE WOODS, I thought I knew birds.

I didn't.

I didn't know Ohio had cormorants or kingfishers.

I didn't know the names of catbirds or towhees. I didn't know the cedar waxwing or the rose-breasted grosbeak. I didn't know the eastern kingbird or most of the migratory warbler species that pass through my state each spring.

My birds, the birds I knew well, were the robin, the mourning dove, the Carolina chickadee, the cardinal, the crow, the nuthatch, and the northern flicker.

I knew eastern bluebirds through stories, but as far as I could recall, I had never seen one in my youth.

They belong to the legendary past of Ohio wildlife, the past in which my dad made summer spending-money by trapping muskrats and selling pelts (yes, that sounds impossibly outdated, but it's true), the past I was told held ubiquitous minks and foxes and wild turkeys, animals I rarely saw during my childhood.

Part of me suspects I must be misremembering, that I must have seen bluebirds, that this hole in my recollections is another symptom of mental pain and dissociation.

Or it may simply be a result of my habitat as a kid.

In my childhood, I was mostly a creature of the woods. I knew the nuthatches and woodpeckers and the sorts of birds that would visit our birdfeeders, but I didn't often encounter species that like open meadowlands like bluebirds and tree swallows. I didn't (and don't) have anything like a comprehensive knowledge of bird species in my area, but I hadn't realized how much my focus on woodlands had influenced my knowledge until I returned to the subject as an adult.

I thought of the eastern bluebird as rare enough to be a thing of myth.

I knew their population had plummeted, and I knew I had never spotted one.

I had glimpsed the bright blue feathers of an indigo bunting once or twice in my parents' yard, but never a bluebird.

The eastern bluebird, the bird that, as naturalist and writer John Burroughs put it, "carries the sky on its back and the earth on its breast," had always been a ghost of the past, another story of my mom's childhood, as magical and absent as the grandmother I would never meet.

And then, one average afternoon, I saw a bluebird.

That spring had been a muddy slog.

The hiking trails.

My therapy.

My ongoing unease about taking medications.

All of it was slippery and boggy and squelched beneath my boots, threatening to snatch them from my feet.

Winter was the sharp, crystalline clarity of pain finally acknowledged.

Spring was the thaw, when everything was exposed and tender and full of the clumsy, awkward power of new growth and sudden expectation.

My walks in nature had become part of my basic self-care, like brushing my teeth, which is to say, I managed them most days.

One day in April, I drove five minutes north of home to Delaware State Park, a 1,700-acre recreation area with a lovely lake and campground. The place was new to me. Another discovery that seemed obvious once I knew it was there.

I parked and walked down a muddy slope to look at the lake.

I saw a kingfisher perched on a tilted, barkless corpse of a pine leaning out over the water. I knew the name, but I didn't know they were in Ohio. I was excited, but I also, once again, felt like a child. Like an absolute novice. I was feeling the friction of a practiced ego bumping up against the awe and humility of encountering the unknown.

The kingfisher ruffled her feathers and departed for the far side of the lake, apparently unhappy with her fishing spot or her new spectator.

I shook my head at the enormity of my own ignorance and turned from the water.

There, perched on a spindly tulip poplar sapling, was an eastern bluebird.

It was ridiculous.

It was like an escapee from a Disney movie, a fairy glimmering in broad daylight.

It was just a bird, sitting there, looking like it was right where it was supposed to be. It *was* right where it was supposed to be.

I pulled my phone from my pocket and took dozens of terrible photos, trying to document the event like a bigfoot sighting, like a two-headed buck had emerged from the woods muttering prophecy.

The bluebird twittered and flew to some honeysuckle to join three more of its kind.

I stared.

Was this important news?

A discovery?

I wanted to call my mother.

I wanted to head straight back to the car to look at my pictures and send the best of them to Leslie, but I decided to finish my walk. I continued my mucky, slippery wander around the lakeshore wondering what I might see next. Probably a pterodactyl or a feathered serpent swallowing the sun.

When I made it home, with muddy boots and spattered jeans, I took to the internet to share my sightings. An acquaintance had suggested several Ohio-based birding groups on social media, but I hadn't gotten around to joining any of them. This seemed like the correct moment.

I joined a group and opened the message board.

One of the first posts I saw was a bluebird sitting on a backyard birdfeeder with fruit and mealworms.

The top comment read, "cute."

Not, "wow!"

Not, "where was this taken!?"

Just, "cute."

I got the message, and further research confirmed that eastern bluebirds were considered common in Ohio. The belted kingfisher was considered common as well.

I hadn't seen anything special.

I had primarily sighted the shape of my own ignorance.

I didn't have big news to share.

All I had done was confirm for myself that I was not as knowledgeable as I once thought.

I had given my fragile new identity as a naturalist a black eye.

Somehow, I had found a way to be excited about birds incorrectly.

I had found a way to embarrass myself while walking alone in the woods.

I had accidentally wandered into a pop quiz, and failed.

I did not celebrate learning something new.

I focused on mourning the fact that I didn't know already.

❖

I went to counseling every Wednesday at 4 p.m.

I carried a little notebook, its first third filled with notes from marketing meetings at Ohio University, its second third rapidly filling with notes about doubts, fears, and troubling thoughts that I had never committed to paper. Ideally, I took notes throughout the week in preparation for my counseling sessions. The notes were partly so that I could recall my state of mind in the days between sessions, partly so I would have something to say when I met with Mark, so that I wouldn't find a way to freeze up and "waste" counseling or to feel useless or uninteresting as a client.

My mind is always ready to invent criteria for failure.

When I first filled out the paperwork for Northwoods Clinic, I listed an inability to write consistently as one of my primary reasons for seeking a counselor. I had several writing projects underway including a novel and a new scripted podcast. I have always enjoyed creative writing, and feeling like I was not giving that passion the attention it deserved was one of my many sources of guilt. Still, in retrospect, it was a bizarre complaint to list on the clinic's intake form.

I was suicidal.

I struggled to get out of bed, to shower, or to accept that pleasure was possible to experience.

Progress on my novel did not merit mention, but I was still struggling to be honest with myself and others. Framing my problem as a work problem, a productivity issue, made it an easier, more respectable topic to explore.

"Yes, hello, my output is suffering. Please help me back on the path to being a worthwhile citizen and worker. A noble goal, we can agree."

So, the writing became a metaphor for everything else, the sadness, the pain, the constant exhaustion, the regret, and the guilt. Both Mark and I knew what we were actually discussing.

"I just can't seem to make myself sit down and write," I said.

"Last time we talked about setting a timer, trying twenty minutes and calling it good enough. Goal met for the day. How did that work?"

I looked out the window and tried to summon honesty. It was difficult. I always wanted to answer, "Actually, I'm fine. How are you today?" I imagined a bluebird flying over the parking lot. A perfectly commonplace bluebird.

"It worked a couple of times. Then I had a bad day and couldn't even manage the twenty minutes. And I guess once I had broken the streak, it just felt like a failure. Like it was pointless and painful to think of. It's not as though I could do much writing in twenty minutes anyway."

"So, some became none. Which then became retroactively pointless. Which then became a new failure on your permanent record. More ammunition for that inner critic."

"Yes. I suppose. I hear myself say these things and it all seems so obvious and easy. Twenty minutes is nothing. Why can't I just make myself do it?"

"On the days you did do the twenty minutes, did you celebrate? Did you give yourself credit?"

"No. Not really. It was just twenty minutes."

Mark furrowed his brow.

"Have we discussed the stages of change?"

"No. I don't think so."

He nodded and then left the room, returning a moment later with a dry erase board.

"Have you considered that maybe you *are* doing it? That, even this week of struggles, you have things to be proud of?"

"Well, my wordcounts and unwritten chapters seem to suggest otherwise."

Mark uncapped a black marker.

"The stages of change."

He drew a simple staircase with five steps and began labeling them, both the steps and the rises between the steps.

Step one: No issue.

Rise one: Pre-contemplation/Denial.

Step two: Acknowledging the problem.

Rise two: Contemplation/Getting ready.

Step three: Ready to change.

Rise three: Preparation/Planning/Research.

Step four: Action.

Step five: Maintenance.

"Sometimes this is drawn as stairs, sometimes as a circle or spiral, but these are most of the main ideas."

"Most?"

Mark turned back to his drawing and drew arcing arrows leading from each step back to previous steps and then labeled the arrows "relapse."

"Relapse doesn't get a step, but it's part of the steps. It's key to the process."

"Okay. I think I get the idea."

"So, which step are you on?"

"Sometimes action. Sometimes preparation, I guess. That action step is slippery."

"So, that's pretty far along this staircase, wouldn't you say?"

"Yeah ..."

"I hear the *but* forming."

"But those first steps don't actually count. They aren't achieving anything."

"Uh huh, except they do. They do count. This is how people change. Does it make sense to ignore the steps you've already taken just because you've taken them?"

I frowned.

Mark circled the stairs and the relapse arrows.

"This is not a linear, straightforward process. It's a messy process. It's a cyclical process. But this is the process, the reality. This is how people change. It doesn't matter that it doesn't seem tidy or efficient. What matters is that it's the honest representation of the process and each of these steps deserve to be respected and acknowledged."

"Alright, but is it really a process if the progress is so nonlinear?"

"Are you asking what the point is when it isn't a perfect straight line from goal to achievement?"

I looked at the floor. Yes. That was what I was asking.

Mark tapped the board.

"This is the process. The process is what it is even though it doesn't always match our expectations for ourselves or the expectations others put on us. But being real about the process is the way forward. Taking one step forward and then two steps backward is not a calamity. It's the inevitable process."

"So we just expect to fail again and again?"

"Is that so bad?"

"It certainly feels bad."

Mark shook his head in the way I recognized to mean "reset and reapproach."

"Think of it in science terms, not in moral terms. Each of these attempts at the next step is an experiment. It doesn't give us victory or defeat. It gives us data. We need that data. When we reframe from 'success or failure' to 'experiment and data,' we capture the truer sense of what *this* is."

He tapped the drawing again.

"This process. This cycle. It's not a judge or a test. It's an ongoing experiment. It's not shooting an arrow at a target, where we hit or we miss. It's more like planting a garden. It's watching the birds. It's being alive and messy. Folks make New Year's resolutions and try to skip from the pre-contemplation step straight to the action step, but that is often a path to disappointment because it doesn't honor this process. These steps are not prescriptive, they are descriptive. All of it, relapses included, are just how people change. The process becomes less painful when we honor the honesty of it, when we honor the truth that nonlinear progress is still progress."

I felt my face getting hot.

This felt so simple and so vitally true.

"So what does honoring this process mean, then?"

Mark tilted his head.

"It means that we don't expect to be perfect. It means that when we miss a goal and slip backward, we smile and acknowledge that it isn't a tragedy, that it is the process working as we expected. When we give ourselves that honest, essential generosity, it isn't so painful to try for that next step, because missing isn't a refutation of what we have accomplished or who we are. It's just another experiment. More data."

I nodded. Sipped my coffee. And, once again, tried not to cry.

Mark wasn't just trying to cheer me up or give me a trite cliché to hold like a paper shield before the onslaught of hopelessness. I sensed no toxic positivity here. I felt the stark reality of what he was saying. He was just telling me a truth, one I needed to hear articulated.

Of course progress with something as complex, fluid, and natural as a human mind was nonlinear. Of course much of the process would be invisible. Of course every effort toward a new way of being is an experiment and even failures deepen our understanding.

Humans are not efficient machines. We are animals. Certainly our subjective meanings, our beliefs and identities and goals, couldn't be reshaped along simple, obvious paths and timelines. Lives are art as much as science. It felt so strange, in that moment of realization, that I would ever have thought differently, that I ever condemned myself for failing to change instantly and predictably.

Even so, I felt a jolt of resistance every time I tried to apply this reasonable, self-evident worldview to my assessment of my own efforts and setbacks. There was a gulf between the logic I felt in the stages of change and my ability to wield that logic in a way that exonerated myself from harsh judgement. Depression defends itself. The status quo defends itself. The stages of change felt true ... for other people.

At the same time, my mental resistance to the lesson, the separation I felt between understanding and actually feeling the stages of change, seemed like a hopeful sign. The hope sprang from my resistance itself, from the feeling that my internal critic was snarling and snapping against the wisdom of this new way to view goals and change. I had found something that my depression perceived as a threat to its armor of shame and regret.

Fighting depression is often like fighting a poltergeist. It can push you, but you can't seem to push it back. The resistance I felt was like finally getting my hands on the ghost. A punch that connected with something solid.

I had been grasping at shadows for months, and now I had something. A weak spot, perhaps.

A missing scale on the dragon's underbelly.

I sat in my car after the appointment, reading and rereading my notes on the stages of change, feeling my vicious condemnation of myself wilt and surge, push and retreat, struggling to shut out this new understanding.

There you are.

You don't like this one, hmm?

My setbacks in writing or mental health or any other long-term struggle were not evidence that I was a failure. They were just evidence that I was a human in a struggle to change. I was experiencing the process. I wasn't failing. I was doing it. I was an imperfect person, just like everyone else, gathering data and moving forward in a circular, organic, unavoidable dance between my desires and my limitations.

It was a powerful kind of understanding, and from it a new thought emerged.

The stages of change, the process, the experiment, were not something to generalize or universalize. I was not learning about everyone and everything. I was learning about me. The change, the struggle, the outcomes, they were all about me. There would not be *a* secret to fighting depression. There would be *my* secret to fighting depression.

There is no central wisdom that turns messy human endeavor into easy, linear efficiency.

I was beginning to understand that I would not find one answer to even my unique struggles. My struggles and goals change. I change. Learning to manage my mental health was not about discovering one skeleton key answer. It was about gathering a set of tools.

There is no final victory to be found over our own natural, beautiful imperfection.

The kind and worthy wisdom we need involves learning to embrace the truth that we will never master ourselves, our lives, or our worlds, though we can learn to view the concept of mastery as what it is: a toxic fantasy that clashes with the reality of our minds and our nature.

We don't need perfection, and we don't need to land upon one final, objective truth.

We need to respect our own unique struggles and our own subjective truths, acknowledging that they will grow and change like the living, natural things they are.

It didn't matter how other sufferers dealt with depression.

It didn't matter how other writers managed their productivity goals or what they counted as success.

It didn't matter how common the people in online groups considered the eastern bluebird.

Yes, my efforts were an experiment that yielded data, but the data is mine. Not universal data. Mine. For me. Handcrafted and intimate. The lessons I learned applied to the subjective universe of my inner life, my unfolding understanding of myself.

It's true that often the data we gather is of use to others because we are all so closely connected, but what use was benchmarking my meaning, my journey, against someone with a different life? A different mind? A different set of goals, challenges, and experiences?

What did it matter if others considered the bluebird a common sight?

It was rare to me.

It was special to me.

The truth is, the person taking the notes is the one who gets to decide what is noteworthy.

The person on the journey to change gets to decide what counts as progress.

The trick, of course, is that deciding I was making progress and that my efforts were worthwhile was, itself, part of my path forward. I had to learn to give myself permission to accept my setbacks as productive. I had to learn to give myself permission to define and celebrate my non-linear progress. I had to come to the understanding that imperfection was not a reason to deny myself these permissions, nor was my habit of defining my shortcomings through comparing myself to an idealized fantasy of my life and others' lives.

Part of the motivation behind this temptation to compare ourselves to others is a desire to make our subjective feelings seem like objective facts, so that we can defend them without claiming any authority or liability in defining our own meaning. So that our milestones and priorities are simply correct. We can, for example, take shelter in accepting a typical conceptualization of success in lieu of recognizing our role in defining success for ourselves. This, once again, is a way to refuse the call to make and respect our own meaning.

We don't need to measure our notions of happiness or progress against imagined universal constants for those concepts. They don't exist. Even the typical notions of success change with time, context, and fashion. We do not need to prove that our ideas of identity or purpose are watertight in some objective, evidence-based sense. That isn't how meaning works. Meaning is a story, an act of creation and interpretation, and for human beings meaning is at least as important as fact.

If you doubt this, ask yourself why wealth (crafted meaning) takes precedence over climate (objective reality). Ask why nations (crafted meaning) take precedence over collective population health (objective reality). Ask why the question "Who are you?" almost never refers to a person's mass or volume. Our understanding is founded on stories.

Success or failure. Victory or defeat.

Fraudulent or legitimate.

These are building materials we may use to construct our identities, but in the end, we are the ones deciding on the shape of who we are.

While this is true, it is not so simple.

At least, it isn't so simple for me.

When constructing my own identity or building a toolset for mental health, the components must *feel* true. That sounds reasonable, but it's often a galling complication in my pursuit of wellbeing. I can't just choose the lessons I want to internalize. The information must smack of authenticity. This is why learning the stages of change felt like a breakthrough. The stages felt true immediately. That feeling is how I knew I had acquired an important tool.

We must believe in the ideas we cobble together into personhood. It's why typical toxic positivity catchphrases like "God won't give you more than you can handle" or "Where there's a will there's a way" feel so utterly useless and patronizing to many of us. We instinctually reject these hollow non-answers. We can sense oversimplifications. We can sense condescending dismissiveness dressed in imitation of wisdom.

The stages of change were vital to me not just because I needed them (which I did), but because they rang true. Mark's explanation of the stages felt rooted in real life, not in pure optimism or wishful thinking. In an odd way, the stages were an antidote to my own poisonous wishful thinking, the wishful thinking version of my life and goals that

I was forever falling short of achieving. I was falling short of a fantasy. I was, once again, plagued by the phantom of an idealized life. The stages of change reoriented me to reality, a place where I could learn to find acceptance for myself freed of unrealistic expectations.

Early in my counseling, I was obsessed with external/objective measures of success, so much so I could not shift toward subjective/internal self-love and acceptance. Truth, as I understood it, had nothing to do with my own decisions or interpretations. Truth came from "out there," from objective fact alone. Yet, the stages of change felt like a breadcrumb trail of objectivity leading me back to my own subjective meaning and power.

My stumbles and setbacks were, objectively, not aberrations. They were expected. They were typical. They provided data. Absent of judgement, my struggles represented the usual process of seeking change. So, how could I, subjectively, count my trials and errors as failures in defiance of the objective model? I couldn't. If the process was truly messy, nonlinear, and different from person to person, mind to mind, the logical thing to do was to determine my own, subjective milestones for success. What else was there?

These ideas worked for me because in the scales of my judgement, they read as honest, earnest, and true. The stages of change were not a cheerful aphorism or a leap of faith. Here was an idea that felt solid, an idea that meant the work of mental health could be done in a way that honored my current sense of what is real. I didn't have to set aside all incredulity in order to engage in CBT. I could feel like a critical, discerning participant in my counseling sessions and still make progress toward managing my depression. This was a revelation. Counseling was not just toxic positivity or open-ended questions aimed at providing a forum for catharsis.

There are many approaches to counseling and many ideas meant to battle mental illness, but I have found that only those with the ring of truth belong in my toolbox. So, as I contemplated the stages of change, I found myself in a position to rebuild and reframe.

Yes, it felt true to me that missteps were data not moral failings. It really, authentically felt true. But that feeling alone was not a solution. It was a call to action, a call to stop mourning my past and begin imagining a kinder version of my future.

Mark sent me home with a few different worksheets about the stages of change.

I sat at my desk, half-reading them.

On my computer monitor was a Google search about eastern bluebird conservation.

I looked down and read a worksheet labeled "The Cycle of Change / Prochaska and DiClemente" that illustrated the steps as an upward spiral that included relapse in its core structure.

I looked up and read about how the bluebird population in Ohio had plummeted by 90 percent due to habitat loss, pesticides, and competition for food and nesting space from invasive species like the house sparrow.

I looked down and read the words "learn from each relapse."

I looked up and read that the bluebird population was rebounding thanks to the efforts of groups like The Ohio Bluebird Society and others, people working together to place and monitor nest boxes, along with other habitat restoring measures like choosing native plants for landscaping. Community science and conservation. Small acts aimed at large change.

I picked up my phone and sent one of my blurry bluebird pictures to my mom.

She responded quickly.

"Wow! Is that a bluebird? We never see them around here!"

I smiled.

"Yes," I replied. "I was so excited to spot it."

❖

Learning the stages of change allowed me to approach the process with much less fear of failure and shame.

Once I could set aside shame, I could begin meeting myself exactly where I was.

I could set aside comparisons with idealized hypothetical versions of my life and seek progress through understanding without burdensome regret.

It was time to reimagine and rebuild my identity, but the first step was accepting the materials I had within reach and the imperfect process of construction.

You can't build until you stop mistrusting your materials and begin thinking of their merits and possibilities.

Dwelling on regret and identifying potential are antithetical impulses. There comes a time when you must claim what you were taught to distrust about yourself. Claim the sources of your shame. Claim the things you don't mention at the dinner table. Claim what harms or haunts you. It doesn't matter that these are not the building materials you would have chosen. What matters is that you have them.

When you stop mourning, you can build.

When you stop fearing arbitrary markers of failure, you can build.

All creations and creators have limits. All humans have limits. You are no different.

But we don't get to learn who you are, who you might become, until you turn away from shame, honor exactly what you have at your fingertips, what you have here and now without wishing for a version of you that never existed, until you honestly catalog your resources and challenges without judgement and choose to build something wonderfully imperfect.

I think we all understand that nobody sets out to build something, whether it's a chair or competency with a musical instrument, and succeeds without first making countless mistakes. We understand that. Why would building identity or contentment be any different? Why would building resistance to depression or a toolset to fight anxiety be any different? Why should home improvement projects within the house of selfhood carry more shame and fear than painting a fence?

Because we are supposed to be perfect?

Says who?

Because we are meant to be practical, productive, and product-focused and changing ourselves is often invisible, abstract, and nonlinear?

Someone else will always be willing to spend your time and choose your purpose for you until you say, "enough."

Depression defends itself.

We can always think of a reason to doubt the process of growth and healing.

Yet, one of the wonderful things about a human mind is that we can reshape it with our choices, intentions, and actions.

We can talk back to doubts.

We can bring reason to bear.

We can learn things like the stages of change that recontextualize our struggles and imbue our setbacks with purpose.

We can take hard leaps into the unfamiliar that become less hard with practice, leaps like acknowledging we are worthy of love and care, leaps like cultivating the idea that our mistakes and stumbles are not moral failings.

The bluebird population was not increased by a grand, perfect gesture or the power of a single willful genius. It was increased by many humble human efforts over time. Think of that. The trajectory of a species shifted by regular people placing, monitoring, and maintaining little wooden nest boxes and choosing plants like native viburnum or wild elderberry over boxwood, Japanese barberry, and forsythia. A seemingly impossible thing accomplished not by some masterstroke of perfection, but by the combined efforts of a bunch of small, limited people sharing a common goal and taking steps to achieve it.

Things happen when we give up judgement and simply build with intention and consistency.

Not flawless intention.

Not perfect consistency.

What you have will be enough.

What you build will be yours.

Your flawed, human process will, in the end, be part of what makes what you built special and uniquely yours. There is nothing wrong with grieving our hardships, but too often I allowed grief, mourning, and shame to coalesce into a fundamental part of my identity.

In winter, I had learned to accept that I was imperfect, that I was ill. In spring, it was time to accept that the process of growing and healing would also be imperfect.

That last sentence looks so obvious, but it felt oceans away from obvious to me. It has gotten easier, but I often still fail to accept my shortcomings, and that, I know, is only natural.

Part of the problem is that I have always cherished the idea that there was some true, pure version of myself and I fail to honor him when I stray from his imagined path.

Except it's just not true.

Perfectly or imperfectly, we are never meant to be just one thing.

In your life, you will be many different people.

Your cells.

Your goals.

Your needs.

The only real "purity of self" is our continued willingness to find meaning in change and embrace the wandering trek of telling the story of ourselves, to ourselves.

Chapter 10

CROW

———————

IT WAS MAY, a month when I would be teased and impersonated by crows.

But that came later.

The crows sitting on my back shed didn't know me well enough to tease me. In fact, they didn't trust me well enough to come within fifty feet, despite my attempted bribes of peanuts on the back porch steps. They ignored my offerings or didn't feel like competing with the blue jays who enthusiastically accepted my hospitality.

These were the cemetery crows, a group of birds I saw most days, at a distance.

They perched on gravestones and statues, picking at chokecherries and somehow accentuating the silence with their conversations.

I watched them, admiring their size, their comradery, their glossy plumage. I wished I could befriend them somehow, that I could join their clique. Leaving aside their macabre associations, there is something otherworldly and arcane about crows. They exude intelligence and secret purpose. A murder of crows is a living library, a wealth of accumulated knowledge enfolded between sable covers, written in a language decidedly separate from human texts.

Also, they're just plain cool. I was already a strange guy who walked the cemetery alone as the sun dipped below the horizon. If I had a gang

———————

of crow pals accompanying me, I felt sure I could really become a legend in my little neighborhood. Alas, the crows didn't care about enhancing my mystique.

I can lose hours watching most any species of bird, but crows are different.

Crows are magic.

I don't just admire them. I feel a kind of metaphysical weight in their presence, a pervasive significance behind their glances.

I longed to be a part of that, but like the birds themselves, their magic stayed firmly out of reach. I could feel it, but I didn't know how to connect that feeling to my own life. I was relegated to the audience, a spectator, not a participant.

Spring grew older, and I began to feel a new kind of strength blooming within me. I was feeling better. My energy was swelling and ripening with the blackberries, but a central question arose again and again.

So what?

What was the point of my recovery and revitalization?

I continued to struggle with suicidal thoughts, but this central "What's the point?" was different. It wasn't despair. It was a specific kind of lack. I don't think "What is it I am doing with this life and why is it worth the effort?" is necessarily an unhealthy question. It's a natural question. It isn't a question that needs to be answered by conventional productivity or achievement, but it needs to be answered by something.

More energy and a clearer mind didn't feel like a destination. They felt like a means to an end, and I had no idea what that end might be. A path was opening to ... something. Having been primarily focused on enduring misery and winning external validation, I was out of practice having internally driven goals or hopes.

I was coming to terms with being an ill person and an imperfect being.

That acceptance, while vital, didn't change the fact that life is temporary and sometimes uncomfortable. The simple fact of my healing didn't give me sudden purpose or direction. It gave me a new set of questions and it gave me options. It gave me a voice, but didn't tell me what to say.

I was feeling the presence of hope and meaning in my walks beneath the trees, in my quiet observations of herons and trilliums flowering on a hillside, but it was often like passing a person on the street. The fact of their presence eased my solitude. I could feel the spark of shared humanity between us, but I couldn't call out their name or ask them for help. Our connection was theoretical. Nature was back in my life, and its presence was a balm, but something was missing.

The lush tangle of summer in Ohio was quickly overtaking the young woods where I walked. The streams and rivers were swollen with rain and punctuated with tadpoles. Everything smelled of sweet, chaotic vitality. It sounded like an orchestra tuning, merging into a cohesive melody, then back to tuning. Possibility. Fruition. Returning to anticipation.

For the first time in many months, I felt I needed to envision a destination beyond my own healing. I was moving beyond stubborn existence. I had weathered years of emotional calamity and now it was time to build. But build what?

The first counselor I visited, the one who told me that maybe I needed to suffer for my art, had also said that I might need to come to terms with the idea that I am too mentally ill to hold a regular, full-time job. While I ultimately rejected her approach and viewpoint, her comments stayed with me. Did I need to accept suffering as a fundamental part of who I am? Was I too ill for employment? I wasn't certain. She had mentioned the word *disability* during our conversations, and I felt uneasy about navigating that term and how it might fit into my sense of self and future plans.

I was, from a young age, comfortable with being labeled *different*, but I did not feel the word *disabled* fit my situation.

Leslie had generously encouraged me to step away from work to heal, shouldering the full financial burden of our shared life in order to make my rare privilege possible. Now that healing was at hand, I felt eager to push myself a bit while trying to avoid my past mistakes of seeking advancement and praise without any real consideration for why I was pursuing them.

I wanted to do something, anything, that required me to get out of the house and interact with other people, if only so I could prove to myself that I still could. I wanted to work, but I felt I had an obligation to be picky about it, to avoid squandering what Leslie had given me. I didn't want to return to academia, nor did I want a job primarily aimed at enriching shareholders. I wanted something mission-driven and, ideally, something aligned with my interests. If possible, I wanted something with an outdoor component as well. Thinking back on my favorite jobs, landscaping and construction surveying stood out by virtue of their physicality and time spent in sunlight.

Surprisingly, I found something that fit my wish list.

Mostly.

I reached out to The Ohio Wildlife Center, the state's largest non-profit wildlife hospital. The center treats patients from more than half of Ohio's eighty-eight counties, caring for over 7,500 animals from nearly 175 species annually. I told them a bit about my background and experience with fundraising, writing, database management, and marketing. I was happy to find that they were interested in working with me, even though they didn't exactly have an open position that fit my background.

The assistant director allowed me to start working as a part-time consultant with the eventual notion to move me into the role of events manager, a position that mostly oversaw rentals of the organization's event space, a large, restored barn from 1891 that sat among the center's twenty-acre, wooded education center. In that capacity, the job involved everything from coordinating with wedding parties on weekend evenings to scrubbing floors and setting up tables.

The work was work, but the people and animals I met were life-changing.

There was a wild turkey who would visit the center and peer in the windows during meetings. A recently rehabilitated barred owl who occasionally landed near me to see if I had a mouse for her as I walked to my car. I met a turkey vulture who would untie your shoelaces. There was a massive snapping turtle I lugged around fundraising events to impress donors (while reminding them to keep their fingers away). And, my favorite of all, two crows who, due to their injuries, were permanent residents of the center as "animal ambassadors."

The first time I met the crows, Percy and Lenore, they said "caw" in a human accent. They said it like a human reading the word *caw* aloud. The tech who was giving me a tour shook her head and said, "They're making fun of us. People say 'caw' to them all day, so they've started impersonating us right back."

It was hilarious and incredible.

They would say "caw" and then cock their heads in a way that seemed to say, "We got you."

I have history with crows. My mom used to feed a group of them and, if she missed a feeding, they would tap on the front door as if to say, "Ahem, aren't you forgetting something?" But being actively teased by a crow? That felt like something new and special. It felt like I was being let in on their magic.

More than once, I had the opportunity to sit inside their enclosure along with the Center's director of education. It was beautiful to just sit in their company, speaking quietly to them as they studied me and played with bits of twig and leaf.

The Center's goal was always to return animals to the wild, and the overwhelming majority of orphaned and injured who came through did not stay on as long-term residents and ambassadors to the public. Yet, those injured creatures who did stay had incredible power to shift worldviews. Raccoons, coyotes, barn owls. Foxes and gray rat snakes. The black vulture who arrived as a downy gray fledgling we nicknamed Potato and soon grew into a magnificent, sable creature with a five-foot wingspan, happy to greet humans as friends.

I watched as people realized that snakes are not slimy and opossums are not vicious.

It was impossible to be unmoved by these encounters. Impossible also to spend time with the Center's residents and not see, yet again, that the distinctions between humans and the rest of our living planet are the thinnest of fictions, as ridiculous as walking inside a wooden house and thinking we've somehow been transported outside of nature.

Percy and Lenore would call in wild crows from the surrounding woods and have conversations with them, sometimes pausing long enough to tease me with human sounds as I walked past before continuing to speak in their native tongue.

The experience was not just touching or special.

It did not simply nudge my worldview in a new direction or deepen my gratitude for the wild.

Those creatures, that place, arrived in my life at a pivotal moment and gifted me with exactly what I needed. They gave me permission to experience reality as a newcomer again, to play, to fully surrender to amazement. They gave me permission to reframe the natural world from a source of awe to something more fundamental and intimate. Not a spectacle or a curiosity, but a pathway to understanding myself. It was a new way to answer the "so what" questions that came with my new healing and energy.

It was transformative.

It was time to make friends with a new kind of meaning.

It was magic.

❖

Science and magical thinking are often framed as oppositional concepts and, at a surface level, this feels fair enough. In many ways they do describe antithetical thought processes. The problem is that people tend to take these distinct, seemingly oppositional frameworks and apply them to a dichotomous moral or philosophical tribalism.

One must choose.

Will I be science-minded or magic-minded?

This tendency shrinks both concepts and stifles the broader spectrum of human understanding.

These paths to understanding are not moral frameworks. They are different tools for different tasks. The friction arises when the wrong tool is used for the wrong task. If we use magical thinking to, for example, approach the problem of what is causing an algae bloom on a lake adjoining farmland, we encounter friction. If we use science to approach the problem of elucidating the beauty of impermanence or the worth of a single moment in a human life, we have friction. Yet, each of these tasks is worthwhile.

When we pour all of our values, our identity, exclusively into one or the other pathway to understanding, we inevitably become reductive and dismissive.

Prioritize magical thinking and the subjective becomes king. Suddenly perception overshadows measurement and evidence. Fine for discussing poetry or our own raison d'être, but not so fine for managing a public health crisis or determining the root cause of a contaminated river system.

The opposite extreme is likewise myopic.

Prioritize science and the objective becomes king, stumbling into the arrogant supposition that human-crafted meanings neither shape nor inform human understanding and reality. This viewpoint leaves in its wake a terrible silence, a shadowed throng of unanswered questions. Why? Because there is no objective meaning in the universe.

Both worldviews ignore huge, vital swaths of the human experience.

Meaning is a human construct for human purposes. Why do we study antibiotics more than the communication pathways of pine trees? Why is so much of mainstream civilization structured to serve money more than nature? Sure, we can study contamination in a river system, or the migratory paths of warblers, but in either case, why do we care to do so? The answers are important and subjective.

Every human who has ever had a perspective on reality did so with a human mind and a human viewpoint. Moreover, each of our individual viewpoints are informed by myriad experiences, prejudices, and values. It's inescapable. It's natural.

It matters.

Our meanings, our perceptions, our non-empirical ways of knowing ourselves and our worlds are not fluff, nor pure entertaining distraction to be dismissed as irrelevant or frivolous.

174

Meaning shapes the world. Literally. And—the vital point—it is created by us.

Facts are important, but for humans, there is no fact that can answer the inevitable question that follows in its own wake, its persistent shadow.

"So what?"

I have met smart, thoughtful folks who approach science as if it were a religion or a viable substitute for any and all subjective meaning-making or magical thinking.

The problem with this is twofold.

Firstly, this approach, by the very nature of science, necessitates a turning away from addressing questions of the ineffable or immeasurable.

What is the point of living?

Why does the forest feel like family this morning?

Who gets to make the rules?

How do I value myself?

Who am I anyway?

What is it about this particular rainstorm that nourishes my hope for the future?

How can I reconcile my love of humanity with my distaste for so many of its acts and creations?

In the end, what was the point of enduring suffering and hardship?

Science is not built to answer such questions, yet they are fundamental to our realities.

Secondly, asking science to stand-in for all consequential human thought and inquiry is a mistreatment of science.

Science is a series of methods for understanding our world through observation, experimentation, and evidence.

It is not meant to stand in for all meaning.

Asking science, "Who am I and why do I matter?" is like trying to pound in a nail with a compass. When the difficulty becomes apparent, the blame does not rest with the compass.

I understand that the appeal of science, of the objective footing in general, is in the evidence. The appeal in evidence is in the knowledge that it exists without our effort or agency, without our need to create, define, or defend it.

Making a discovery and illustrating it with a reproducible experiment. Having your discovery confirmed by your peers. Bringing a measurable, observable fact or phenomenon to light. It is certainly a wonderful and worthy endeavor. It seems tidy and correct. It seems to stand beyond doubt (until or unless new evidence is presented to refute it ... good science is leery of absolute certainty).

Science is a beautiful, uplifting, enlightening process that has ushered a great deal of good into the world, and it is very tempting to pretend that it's the only process we need.

I walk through the large cemetery near my home several times a week and read the gravestones of children lost to diseases that have now been eradicated by vaccines and other marvels of science.

Science is worthy, but it is not comprehensive. It doesn't need to be. Here again, we must push back against all-or-nothing thinking.

I love *Moby Dick*, but that doesn't mean all other books have lost their virtue.

The science-minded problem only arises when we lazily declare, "Ah, this tool seems good, so it must be the only tool we need."

I can see the allure of applying a system founded on objectivity to meaning.

Objectivity seems to justify itself. Easy.

Subjective meaning requires our participation. Not so easy.

People are terrified to make meaning and then stand behind it.

This is clear in the way so many of us tie ourselves in knots to make things like money and political power features of the natural order rather than ideas that we collectively make a reality.

What we sometimes forget is that our own meaning doesn't need defending because it's not for everyone. It doesn't need to be for everyone. Your meaning is for you. Your mind is quite literally creating your reality. The entirety of the universe, as you perceive it, is founded on the form and function of your brain. It's absolutely wonderful that we can connect with each other on shared reference points of objective fact and communal meaning. Wonderful, but not the best and last word in the question of what kind of knowledge matters most in a human life.

I have found that the concept of magic serves my needs well.

Magic.

Crafting meaning, in concert with our natural selves and the natural world, in a way that shapes our realities.

Magic.

Giving shape to the ineffable perception that our living world is more than the sum of its individual, physical parts.

Magic.

Intentionally imbuing oneself and one's own meaning with weight, power, and influence.

Magic.

Owning the power to stop thinking of a meeting with a crow as nice and start thinking of it as holy.

I don't doubt that some would disagree with my very broad definitions of magic, but I stand by them. They are, after all, for me. My meaning. I use the word *magic* because it has the correct connotations for the sort of meaning I want to define and communicate, to you and to myself—reality-shaping, personally crafted pathways that change the nature of how we understand our world and our places within it. Magic. A word that marries our outer and inner worlds, marries our physical senses with our beliefs and feelings.

I adore science.

I also love the concept of magic.

I don't find that these ideas conflict.

Fact and magic go together like sound and song, like ink and poetry, like truth and metaphor.

A good metaphor doesn't destroy the truth.

It brings it home to a personal context.

For me, magic serves a similar purpose. An interpretive act in which the interpreter is as much a part of the equation as the interpreted, a celebration of the role our unique perspectives have to play in our lives, in our understanding.

My sort of magic involves a willful cultivation of awe and gratitude, but I don't need my sort of magic to be the only "correct" sort. I don't need to dress up the subjective in the clothes of the objective.

Again and again, I have found that ego is not a friend to understanding. The timid need to be unassailably and universally correct from the outset can shut down our ability to learn and to be enriched by diverse viewpoints.

All magic requires our intention, our choice to participate in an open and vulnerable spirit. We must choose to meet it halfway. And when we do, we find that magic isn't about turning away from reality. It's a synthesis of it, the wheat of experience crafted into the bread of meaning. A nod to the worthy but unquantifiable virtue of our consciousness and the miraculous world in which we find ourselves.

Metaphor. Meaning. Magic.

These things don't simply describe reality.

They change reality in meaningful ways.

The change is largely within us and cannot be measured—we are the place in which our realities are created and interpreted.

There are many ways to dismiss such magical thinking through a blunt misuse of science that shrinks rather than enlarges our human experience.

I have heard "love is just chemicals" spoken in a tone of dismissal.

Yeah?

Okay.

So is the churning inferno of the sun.

So is the bedrock of the earth.

So is the living fountain of a blooming cherry tree.

The problem with "love is just chemicals" is not the word *chemicals*. It's the word *just*. It's the presumption that naming or describing a thing closes the subject entirely, reduces it down to the confines of the label.

Words matter. They themselves are potent magic. As such, it's worthwhile noticing what our word choice is doing to our realities. Are our word choices closing gates or opening ways to larger vistas?

If you need to call upon an old word like *magic* to fully appreciate the beauty of love, of all that which is vivid and real, do so. I certainly do. Truth and fact are sisters, not twins. Attempting to confine human experience to "just the facts" sacrifices being free and whole in order to sidestep our bigger, slipperier questions.

The thing is, we all step away from the data. We must. The data is silent on so many important issues.

The paved road of facts tapers to a dirt track beneath our feet, a muddy deer path through the brambles. The trees lean in.

"Life is short and uncomfortable. Why bother putting on socks? Why make the coffee? Why endure one more staff meeting?"

Science can't answer this for you. Not without your help.

A *what* is not a *why*.

A *how* is not a *why*.

Whatever you answer, it will be correct.

Can you feel the weight of that? The power of your own homemade meaning?

Be careful.

Be thoughtful.

Answer kindly.

What you decide matters.

What you decide is your own kind of magic.

Our identities and our whys are subjective.

A scale map of the mountain is objective.

Yet the map is not the territory.

People use subjective and objective as if there is no commerce between them, but there is. This is a central component of my understanding of magic. Magic is not hurling thunderbolts or mind control, but nor is it pure imagination or restricted to change for and within the self. The way we create meaning shapes the world.

For my purposes, magic is philosophical. But for my purposes, the fabric of human reality is as much philosophy as it is blood and bone and sun and stone. This isn't much of a leap for me, a person whose worldview has oftentimes threatened to be quite literally lethal.

Magic is not just fantasy. Believing such things requires a mix of earnestness and playfulness that is sometimes difficult to hold in balance. Yet, I have seen a crow be simultaneously somber and playful. I have seen a sugar maple stretch branches across the past and present, memory and imagination. I have seen blooms in winter and death in springtime. Nature embraces the project of holding opposing forces in graceful harmony, so why should I turn away from my own version of the task?

❖

The crows taught me something about giving myself permission to feel wonder, to respect myself more by taking myself less seriously, by stepping away from a *just the facts* mentality. I did not need objective confirmation that my meaning was correct any more than I needed permission to be awed by the sight of a bluebird. I did not need authorization to plant a wildflower garden within my own perspective.

That May, I started writing poetry again.

I had been struggling to write, afraid that it was wasted effort when my energy levels were already sapped by depression.

I realized that much of the poetry I had written over the last decade or so felt strenuous. Dense, academic, trying-too-hard lines aimed at communicating sophistication as much as sharing ideas.

I wanted to do something different. I wanted to break in order to create. I wanted to give up trying to be impressive or correct.

I learned that play is an essential part of my creativity. Perhaps more importantly, I realized that if you can fail doing it, then it doesn't count as play.

I wanted to give new shape to the magic I was allowing myself to feel. This wasn't the magic of esoteric knowledge or daunting complexity. This was the magic of campfire sparks or finding a wild turkey feather. The magic of unexamined joy and play. Capturing that magic meant writing something easy and welcoming, a simple stroll through the trees. I wanted to write thank-you notes to nature and try to map out the intersections between personal meaning and our beautiful, tangible world.

Compared to the poetry of my academic training, I was writing what I thought of as aggressively accessible poetry, poetry aimed at connection and communication more than artistry. And, to my surprise and delight, I found a receptive audience on social media, a community that grew quickly and gave me a new sense that I had useful things to share.

I allowed myself to build a frankly confusing online presence under the name The CryptoNaturalist. It was the title of my strange indie podcast about fictional nature—real love for pretend creatures, a metaphor for the magic I felt just beneath the surface of our visible world. In retrospect, the premise of my podcast arose from giving myself permission to place enthusiasm well out in front of expertise. I suspect I was worried I didn't know enough to discuss real nature, so I preemptively removed fact from the equation entirely. Later, with the increasing prevalence of cryptocurrency, I would regret the confusion sometimes caused by the first six letters of my online name. Still, I cherish a slim hope that one day *crypto* will go back to being just an evocative prefix meaning "hidden." I probably shouldn't hold my breath.

One of the readers who enjoyed my poetry sent me a message with a picture of a crow. She apologized that it wasn't a very unusual bird, but she thought crows were beautiful and fascinating. She went on to say that she wanted to get out and hike, but she really wouldn't know what she was looking at once she was out there.

In short, she had a desire to go to the woods and be among nature, but she didn't think she knew enough to do it correctly. She didn't think she had permission. I could relate. Visiting nature, in her case, felt somehow suspect and fraudulent. She doubted her own ability to make meaning from a walk beneath the trees.

I read her message and felt a lump in my throat.

I understood what she meant and I grieved for her.

She felt she wasn't allowed to be in nature because she didn't know the right words for what she might see. She thought she might embarrass herself.

I wrote back.

I told her that, even though I thought of myself as a nature-lover, I doubted I could name even a third of the species I saw in any given patch of my local woodlands, even sticking to common names. I told her that crows are beautiful beyond words, that nature is populated by grand, ancient, gorgeous wonders, whereas our language is populated by small, finite, relatively new collections of sounds and letters.

We don't require words to appreciate grand wonders. Wonders older than the language with which we converse are not limited by our names for them. I told her that experts are important, but so too are enthusiasts. And, a bit shy about my own vulnerability, I told her about magic, about my sense that there is more to feel beneath the trees than pride in one's knowledge of names or even our admiration for aesthetic loveliness.

She sent back a single heart and "thank you."

I am not sure if she went on her hike, but I hope she did.

Words can be walls, but they can also detonate them into showers of shattered brick.

❖

My newfound love of the word *magic* was the final healing tool I gained from that potent spring. I didn't much care to seek out other peoples' interpretations of magic. I wanted to expand my definition of the term by listening within myself and listening to nature. I wasn't looking for a religion or membership in any established school of thought. I simply wanted a name for the wellspring of truth I found bubbling from the rocks inside my mind once I had cleared away the debris.

The crow is special to me.

The word *magic* works for me.

Once we let go of our modern tendency to privilege the objective, "to me" and "for me" lose their stigma as frivolous or irrational.

I need you to hold that idea in your mind so that I can emphasize that magic works *for me*. It could work for you too, or you may find a word better suited or create a bespoke one.

Magic.

That word helps me give power, shape, and respect to the meaning I perceive in the world. Others will need different words.

Values.

Faith.

Religion.

Philosophy.

Practice.

Purpose.

All seem good and proper to me, with one very important caveat.

These subjective meanings can damage us and our world when we, weak and cowardly, need our own vital, subjective truths to become universal, to become objective facts, only real and worthy if we see them mirrored in the lives of others. Then, we get tyrants and orthodoxy. We get zealots and blinding obsession with the project of making our, capital T, Truth as objective as a mountain through the use of violence or shame.

Too many people are afraid to recognize that human meaning is made by humans and are afraid to take ownership of that power and responsibility.

Too many are afraid to live with the possibility of change that looms when we accept that our created meaning is both real and fluid. The meaning is inside us. It changes as we change. Our understanding of reality and the meaning we derive from that understanding will shift and evolve over our lifetimes, which does nothing to lessen its essential value or potency.

Meaning doesn't ripen out in the fields.

It doesn't swell into sweetness beneath the sun.

You won't pluck it from a branch, won't find it tumbled whole and waiting in the grass.

It ripens inside your skull, in the light of your needs and intention.

What you crave is uniquely yours.

Look inward.

For me, these ideas coalesced into "magic."

In our own mental landscapes, spring may come when we need it. Meaning may grow where we plant it. Meaning, like nature, is generous. Build it a trellis and watch it climb. See it sprouting in surprising places. In a note from a stranger. In the return of an old passion. In the call of a crow.

SUMMER

SUMMER IS WHEN Ohio tries to sing with its mouth full.

Nectar and humidity.

Even the air has stuffed pockets.

The table is heaped with hooked burrs and sweet berries.

Plenty upon plenty.

Everything is alive with want, stained with it like purple fingers plucking mulberries.

Plants wrestling for every crumb of sunlight beneath the canopy, deer flies mining for ruby drops of blood, the same wordless need glowing like an ember in the unifying heart of the crowd.

More.

More growth. More light. More taste. More life.

Abundance, when it is here, has a way of feeling like the rule rather than the exception. A banquet obscuring the bare table beneath, obscuring the idea that it was ever bare.

That it might be bare again.

Chapter 11

GRAY RATSNAKE

I KNEW THE GRAY RATSNAKE must have been nearly six feet long when I spotted her from my car.

She was on a busy roadside a stone's throw from a four-lane highway pulsing with rush hour traffic.

The traffic was particularly bad that June afternoon, and I was wishing I had remembered rush hour before choosing that moment to drive to Shale Hollow for my walk.

I turned off State Route 23 and there was the snake, nose nearly to the pavement, the foremost portion of her body retracted in an S-curve as if she had just been startled by a passing car.

I clicked on my hazard lights and pulled over before making any conscious decision.

"What am I doing?" I asked nobody.

But I knew what.

That snake was headed toward a busy highway, and I couldn't ignore her.

I started to exit the car and paused.

There was a jogger passing, footfalls slapping along to the beat of his earbuds.

There was a nearby mini-golf full of laughing families.

An old, familiar voice spoke up.

You're about to embarrass yourself.

You're about to look ridiculous.

The voice wanted me to shrink down to invisibility or, at the very least, keep driving and forget I saw the snake.

Keep your head down.

Fit in.

There are times, more times than I would like to admit, when I would have listened to those voices and driven directly home, ashamed of myself both for leaving and for nearly staying.

Not that day.

Counseling had helped.

My own voice, the one that aligned with my actual desires, was growing strong beneath the summer sun.

I began to have answers for my inner critic, answers like, *It's really none of my concern what those people think.*

A pickup truck shot past blaring its horn.

I got out and approached the snake.

She was gorgeous.

Over five feet of smooth, black scales and snowy underbelly.

"Do you need a hand?"

A forked tongue flicked, tasting the air.

I was certain she was a gray ratsnake, my favorite Ohio species of reptile (depending on the day).

They are sometimes called pilot snakes because they're occasionally found in dens with venomous rattlers and copperheads, and some believe that the harmless ratsnakes have guided their more dangerous cousins to shelter, piloting them to safe refuge in colder weather. It's not technically true, as far as we know, but I do love when a story becomes a name.

Beyond recognizing a favorite reptile, I had made it my business to have basic knowledge of common Ohio snakes, at least enough to recognize our few venomous species.

When I was eleven, I accidentally picked up a timber rattlesnake.

Well, okay, I picked it up on purpose.

The accident was in misidentifying the species before reaching for it.

I saw a snout poking out of the leaf litter and, with memories full of garter snakes and northern water snakes, I gleefully grabbed it just behind the head.

I have always been enthusiastic about reptiles, and snakes were rare enough in the woods and ponds where I roamed to qualify as celebrities. It wasn't until I had pulled the rattlesnake free of the leaves that I realized what I had done. I gave the creature a gentle toss in one direction and ran the opposite way, although I knew full well that I have never in my life been chased by a snake. They have no interest in chasing humans.

From that day forth, I made sure I knew what species I was seeing before I approached. As an adult I also, as a general rule, made a point of not bothering or touching wild animals for my own amusement. Yet, I felt this case was an exception. The ratsnake was headed for danger.

She watched me with surprising calm as I approached.

A second jogger trotted past, not even sparing a glance in my direction.

I was worried that I would agitate the creature or, worse of all, scare her into traffic.

I tried to think calm, deliberate thoughts as I bent and grasped the snake behind the head so she couldn't turn and bite. The animal tensed and I slipped my other hand beneath her midsection for support. I stood and was surprised to feel a kind of gentle curiosity from the snake in my hands, a wait-and-see sort of feeling.

I have a fair amount of experience handling snakes, having had several pet snakes in the past, along with a lifetime of being the weird friend/cousin you call to remove an unwanted serpent from a garage or deck.

If you've never held a large snake, it's a unique sensation. They are dry and smooth, with a surprising heft. There's something amazingly contradictory in the feel of a snake. They can be soft or rough depending on the direction you slide your fingertips. They can be loose and supple, then, in an instant, become like living hardwood, creatures of pure, adamant muscle and bone.

A frightened or upset snake is not a subtle thing, and yet they hold no ill will toward us and some seem quite content to rest skin to skin, taking comfort in the warmth of our touch.

Ratsnakes generally have a reputation for gentle dispositions, and this was far from the first I had ever held, but I expected an attempted bite or, at the very least, to be musked (a foul-smelling discharge some snakes release when threatened).

This snake did not feel upset. This snake felt calm and inquisitive.

She flicked her tongue and relaxed.

I risked doing something foolish and let go of my grip behind her head, changing to an open hand. She was in my arms and was not likely to escape into the road, but I was risking a painful bite. The ratsnake pointed her snout up my bicep and began climbing toward my shoulders.

"Okay, then," I said. "Welcome aboard."

I laughed a nervous little laugh.

I looked around. I was a short drive from the entrance to Shale Hollow Park and there were no good options for relocating her to safety within easy walking distance.

"Well, let's stretch your patience a litter further, okay?"

Slowly, carefully, I climbed back into my car, the large snake's cool skin sliding across the back of my neck.

I thought about my breathing. I thought about projecting calm, friendly thoughts. I thought about magic.

I continued my drive toward Shale Hollow.

I saw a tongue flick in my peripheral vision and wondered what smells defined my car for my visitor. The coffee in the cup holder. My muddy boots in the back. Fabric softener. Beeswax lip balm.

"Like a Mountain" by Timber Timbre pulsed quietly from the speakers as I continued willing my friendly intentions to fill the car along with the music.

To be clear, moving wildlife is almost always a bad idea.

Transporting an animal out of its home territory can be lethal to the animal (and illegal) depending on the species and a variety of circumstances, but Shale Hollow was just around the corner, and this snake seemed to be in immediate danger.

Typically, if you do want to escort a reptile across a busy road, you simply move it to the other side and send it on its way. You move it in the direction it was already headed. You respect that it knows where it wants to go.

For example, I've carried eastern box turtles across busy country roads, helping them to safety.

At the Wildlife Center, one staff member talked about coaxing a large snapping turtle to latch onto a branch and then dragging it across a highway to the pond beyond.

Unfortunately, I couldn't just help this snake cross five lanes of traffic only to reach an expanse of parking lots and shopping centers.

So, I took a chance.

I was feeling fortunate and flattered by the ratsnake's unlikely, instant trust. I was also feeling fairly capable and savvy—until the creature decided to leave me behind entirely and start exploring my backseat. My confidence evaporated. I considering the many, many nooks and crannies in my car from which I would have zero chance of extracting a nervous snake.

"Brilliant work, Jarod. Just brilliant."

I sped up.

Thankfully, I didn't have far to go, but I knew that even a large snake could disappear like smoke in the wind when they wished. I'd seen it happen many times. I pulled into Shale Hollow and kept one hope held steadily in my hands like an overfull soup bowl:

Please be there when I turn around.

❖

My mental illness had changed.

It was not gone or defeated, but it felt fundamentally different.

Hopelessness, depression, anxiety, suicidal thoughts, and insomnia.

These were still potent forces in my life, but they had stopped being so adept at crowding out all other thoughts. I was learning to see their manipulations, distortions, and overreaches. I was learning to hear their shouts and threats and cruel whispers and reply, "I hear you, but I am choosing a more pleasant viewpoint today." Or, "I hear you, but you don't get to determine my actions or outlook in this moment."

The process was neither perfect nor always effective.

Of course it wasn't.

Yet, the fact that it could work, that I was able to outfox my old devils from time to time, changed how I viewed the monsters lurking behind my eyes. I was no longer a prey animal trying just to hide and survive. Survival was no longer my highest ambition. I was armed and on equal footing. I could fight back. I could make new plans and goals. The monsters were still monsters, but they were no longer unopposed terrors of my inner wilderness.

I hurt, but hurt didn't define me.

I learned to reject hurting as an identity or as incontrovertible evidence of a failure of self.

I had gained some tools.

My response to medication helped me conceptualize my pain and fatigue as an illness, not a part of my character.

CBT changed how I thought about thinking.

I had new kinds of handcrafted meaning and the means to build more.

I felt less like a passive presence enduring an uncomfortable life. As I felt my agency grow and strengthen, I realized I had things to do.

Not hateful obligations.

Things I actually wanted to do.

Options.

I had a mind full of old furniture and a powerful need to do some remodeling.

Have you ever walked into a room in your home and thought, "Why these things?"

Suddenly, a question mark hovers above each of your possessions.

Why is it so cluttered in here? Why these books? Why this old chair? How long have those clothes been piled there? Have I ever worn any of them? When was the last time I touched that exercise bike? Why do I allow it to stay? Is there a reason I am keeping this slightly bent pizza pan I salvaged from the trash in college a decade ago? Why are all these doors and windows barricaded against monsters?

I thought about the embarrassment I felt stopping to rescue the ratsnake and realized that, in this time of learning and reassessment, I could question more than just the actively harmful parts of my thought patterns.

I could take another look at less obvious patterns that were not serving me. Why was I embarrassed to be seen caring about a snake? What was the potential consequence?

The answer came easily.

I was embarrassed to be seen caring about anything. Embarrassed and afraid.

It wasn't the first time I had noticed this, but it was the first time I felt in a position to do anything about it. CBT had taught me that thoughts, feelings, and actions were not just found things that must be accepted as they are. I could question them. I could challenge and change them.

Beliefs are often more complicated and enduring, but thoughts?

You can have a conversation with your thoughts.

I considered my embarrassment about the ratsnake and realized it was related to a broad fear of sincerity and vulnerability. I thought of other ways that fear of vulnerability has shaped the patterns of my behavior, past and present.

I felt a spark of déjà vu.

It was a past insight I had unearthed and reburied because I was not ready to face it.

As a student, I cared deeply about writing, but I had a habit of waiting until the last possible moment to write an essay or research paper. It made my life needlessly difficult and, in an unusual moment of honest self-reflection for undergraduate me, after a painful failure in a class I cared about, I tried to write some notes about my motivations for constant procrastination.

I arrived at an uncomfortable answer.

Fear of trying my best.

I understood, even then, that trying is a kind of vulnerability.

If, for example, I made an earnest plan for writing an essay, giving myself the space, resources, and time to do my best work, it would then be possible for my professor to access, and therefore evaluate, an authentic version of me, of my abilities.

This was an uncomfortable prospect. It moved my achievement-focused self-worth too far outside my direct control.

If, on the other hand, I waited until the last possible moment and pulled an all-nighter to write the paper, I seemed to gain in two ways.

If the paper was a success and I earned an A, well then I am a genius because even my fractional efforts were superior.

If I did poorly, good news, I had an ironclad excuse.

"Well, sure, I got a C, but I didn't really try."

So, writing the paper the night before provided an opportunity to be a genius and a reasonable defense from any criticism.

A win win.

Except, of course, the entire process is a way to avoid being a whole person. It is a way of hiding in the shadows while holding a paper puppet into the light. Yes, the puppet absorbs criticism without feeling, but it also absorbs praise without feeling. It doesn't feel at all.

Then and now, I was afraid of earnestness and the related vulnerability.

I didn't want to be seen rescuing the snake because I actually deeply cared about that action. To me, revealing my true feelings and acting upon them is like being nude in public, exposed and unprotected. Showing my real face, not just the paper puppet, means that any criticism tossed my way will land squarely on my jaw. I would have to give up my rhetorical armor.

Those are the apparent downsides to earnestness.

But working at the wildlife center, publishing a podcast, and posting my poetry online were starting to show me the upsides of earnestness.

When, for example, a stranger complimented a poem that expressed my true, unfiltered perspective, that praise was hitting home in a new and powerful way. They weren't praising the puppet, they were praising me. They weren't connecting with the puppet, they were connecting with me.

Earnestness.

The vulnerability of honesty.

As I examined the furniture of my personality, I found I had a name for the object I needed to carry to the curb.

Ironic detachment.

Numb, smirking aloofness in place of personality.

It was a massive old chair, piled with laundry, crumpled papers, and unread mail, that had been in the corner so long I had stopped seeing it. A landmark in my living space since I was in middle school.

I was a child of the 1990s. Sincerity was not cool. Irony was the height of humor.

The problem with any mask is that, if used consistently enough, it stops being a disguise and becomes a uniform. Wear the uniform long enough and it becomes identity. The danger in trying to hide who you are is that you'll succeed and you'll start to see a stranger in the mirror of other people. Do that enough, and you become a spectator, a phantom hovering somewhere in the background of your own life.

I had taken a leap of faith when I made room for mental health treatment.

I was taking another leap when I started to apply what I was learning to other aspects of my identity. Painful, awkward, unfiltered honesty was the only way that counseling worked for me. The ugly truth is, I hated and feared weakness. Part of counseling was admitting my own weakness again and again and again in front of, essentially, a stranger.

Initially, it felt impossible.

It got easier.

I began to understand the benefits.

My risks were rewarded.

Sincerity feels frightening because it is powerful.

Irony and detachment feel less frightening because they are ... less.

❖

I parked by the creek and looked over my shoulder.

Instant relief.

There was the ratsnake, a dark figure-eight on my backseat, a sleek infinity sign tasting the air and looking comically at home on the gray upholstery, flanked by rumpled sweatshirts, chin resting on a well-creased copy of a Jeff VanderMeer novel.

"Stay right there, please."

I climbed out of the car as slow as a sleepless night and did not risk shutting my door.

Somehow, the snake's patience held as I lifted her from the backseat and walked toward the liquid purr of the shallow stream.

We made it.

No bites and no musk.

I laid her gently in threadbare leaves from last autumn and she tolerated me taking a few pictures, a keepsake to show Leslie, before racing off along the creekbank in the direction of the place where I'd spotted the wading heron on that gray January day when the year was new.

The sudden life and speed of the transported ratsnake brought to mind the squirrel I had carried from the road in Athens. A gray squirrel and a ratsnake, two oft-maligned, beneficial creatures. Depression and vulnerable sincerity, two defining struggles of my life. I'd felt considerably less awkward and uncertain while moving the ratsnake than when I'd moved the squirrel, because I better knew my purpose and the creature I was aiding. What else might grow easier with time and experience?

I waited a moment, then walked in the same direction as the snake.

I knew I wouldn't spot her again. That was as it should be. It didn't matter.

I was out in the woods and walking, the old medicine that had brought me this far with the help of many of my fellow inhabitants of this unlikely planet.

I pointed my feet toward the memory of a tall, somber bird and smiled.

The pilot snake guided me in the right direction.

Chapter 12

FIELDMOUSE

"YOU CAN STEP ON THAT MOUSE if you want," my dad said in a conversational tone to my brother DJ and me.

DJ was thirteen.

I was eleven.

The three of us were walking the edge of the woods, surveying the scraps of plywood and vinyl siding that were left over from the shed we'd built on the edge of our property in Newark, Ohio. A muddy blur of a mouse, startled by our passing, darted out from under one board and hid beneath another.

Hidden, but far from safe.

"Just don't tell your mother," Dad added as an afterthought.

My dad's tone suggested two things.

The first was that he was giving us permission to stomp the guts out of the mouse, because he figured that my brother and I wanted to and the only reason we hesitated was a sense that we needed permission. The second was that crushing mice fell into the vast ocean of fun, semi-secret activities that the women in our lives just couldn't understand.

Man stuff.

These two suggestions communicated a wealth of information to me about my dad's expectations—and in my white, middle-class Midwest, young boys become their parents'/teachers'/peers' expectations or they themselves are crushed.

My dad is a good man.

He's a thoughtful man. A caring, loving, and supportive man. A brave man who has, over the four decades I have known him, been an inspiration to me in both his willingness and courage to learn and change, and also in his dauntless dedication to his family.

His version of good is the strenuous kind.

We always stopped to help stranded drivers, often pushing a dead car to safety.

I've heard stories (told by strangers) of him paying other families' rent.

Dad owned a business in downtown Newark, and I worked for him, on and off, as I grew up and found my footing in the world. We couldn't go out to eat without somebody visiting our table to shake his hand and chat. I've been stopped by both county employees and random acquaintances in Newark just to be told what a good man my dad is and how lucky I am to be his son, people who needed nothing at all from me.

He is old-school tough and has grown at least as much in his own journey of feeling and understanding as I have.

While my dad is not defined by this fieldmouse story, the story is true, and I think it's significant that among all my many lost memories, this one survives.

That's what makes it important.

Not because a fieldmouse is a big, memorable wildlife encounter, but because it's not.

An Ohio fieldmouse is the most insectile mammal I know. It's the eyes. Large and liquid black, bulging beads of ink.

When I think of mice, it's hard for me not to think of their importance primarily as prey, as a seminal link in the food chain. When I think of mice, I think of ratsnakes. I think of barn owls and foxes. Weasels and kestrels.

I think of fieldmice and I think of death, white bones, precise as watch parts, gleaming in an owl pellet at the base of a hickory. I think of the numb, unpleasant chore of emptying the kitchen snap-traps while trying not to look at their little whiskers and think "puppy."

I think of mice and I think of the phrase, "You can step on that mouse if you want."

Stepping on a mouse in the woods is purposeless cruelty.

But was my dad wrong that it was something I might do?

I have told you that I loved nature even as a child, which is certainly true, but how I express that love has changed a great deal over the years.

The truth is, I have killed many dozens of animals. Perhaps more. I have trapped, skinned, and eaten squirrels and rabbits. I have illegally hunted songbirds and squirrels with air rifles. I was once tasked with helping control a groundhog population with homemade snares, metal traps, and a .22.

Was I the kind of boy who might have stepped on a mouse given the right circumstances?

I'm afraid I was.

Even though I sometimes caught fieldmice in bucket traps and kept them as pets, carefully tended and fed, it is not out of the question that I might have also killed them on a whim. Mice could be friends, but their deaths could also be small tastes of power, a small way to be a killer. A predator. A man.

I, like my dad, was and am a paradox.

My dad was trained as a welder. He grew up trapping and skinning muskrats for extra money. He lost his class ring by tossing it to someone to hold before a bar fight, one of the times someone broke his nose. I've heard him call himself a redneck.

Years after DJ came out to my parents, he told me that he thought they might disown him for being gay.

My redneck dad did not disown my brother.

When my grandpa expressed bigoted religious views on homosexuality, my dad told him he was no longer welcome at holidays.

My redneck, sports-loving, small-town Ohio dad never waivered in his support for my poetry or his enthusiasm for my writing and publishing, though I don't recall ever seeing him read for pleasure until he was retired and I got him hooked on contemporary fantasy novels.

My dad is not one thing. Nobody is. One of the core dangers of the toxic form of masculinity rampant in our culture is that it smothers difference and distinction. It flattens us into tropes and caricatures. It mandates adherence to a narrow orthodoxy of acceptable masculine traits and behaviors. It offers a kind of blunt confidence in belonging, a shelter from ambiguity, and takes in return nuance, feeling, and real freedom for exploring, defining, and expressing our own identities.

Toxic masculinity diminishes through its insistence on conformity and its rejection of the richness of other perspectives. It's a shackle disguised as armor. It shuns depth of feeling and complexity in personhood in exchange for the paltry prize of silencing tough questions from within and without.

I don't believe masculinity itself is toxic.

Masculinity is part of who I am. It's linked to my sense of self and interwoven with both my mental health struggles and successes. The toxicity arises with the pull toward conformity—more specifically the harmful, often self-defeating, petty, predatory arrogance that is so often sold as strength and uncritically accepted as part of our foundational masculine archetypes.

I found that reexamining my understanding of my own masculinity was part of the work required to move forward with my healing.

Like the shame of my depression or my fear of seeking help and vulnerable earnestness, I had internalized the more rigid, thoughtless aspects of mainstream masculinity long before I understood that I had any role to play in determining my own values and identity. When I began examining the role of masculinity in my own inner landscape, with the help of CBT, I came to understand that not everything I once accepted as part of being a man served me or actually reflected my understanding of the strength and resilience I associated with masculine virtue. It was time to reevaluate.

❖

We didn't step on the mouse that day, but it wasn't because we held some unshakable notion of the value of the mouse's life.

We just weren't in the mood.

That was the nature of my own inconsistent relationship with wildlife, with killing, and with masculinity when I was a boy. I loved animals. Yet their deaths were also a way for me to connect with something that felt primal and, yes, masculine. At least, what I understood to be masculine at the time.

As a kid, books like *My Side of the Mountain* and *Hatchet*, along with a general discomfort with crowds and civilization, inspired me to learn many of the skills needed to live off the land. I learned to trap, forage, make shelters, and practice striking sparks with flint and steel. I was internalizing ideals of rugged individualism and all the things which seem related but are not.

Thanks to all my study and practice, if you dropped me off alone in the mountains ... well ... I might starve a little slower than some other folks if I didn't freeze to death first.

Through media, friends, and role models, I was learning that men are self-reliant and tough. I was also learning that being tough meant men are also often mean, domineering, and numb to empathy or remorse. I thought I was earning myself the skills to leave civilization one day, but those skills dovetailed in unexamined ways with other, darker infatuations with escape that were coalescing around my creeping young depression.

I was learning through countless subtle examples that thoughtless cruelty was one of the principal ways I was expected to perform my masculinity. The insidious part is the difficulty in separating the practical from the malignant, the useful from the toxic, the grounded from the arbitrary.

As a boy, all the elements I sensed were related to masculinity seemed connected and interdependent. Somehow, stepping on a mouse seemed to exercise the same muscle as hunting to eat, bicep strength, my rifle aim, sleeping on the ground, or practicing wrestling skills.

What does being strong have to do with being cruel?

Nothing.

But I couldn't quite parse that fact at the time.

This is the hidden snare of toxic masculinity, the reason it is pervasive and insidious. It wears the clothing of strength and logic.

Breaking the constraints of toxic masculinity is part of my work toward mental health. It helps to realize that the term is frequently misunderstood or used as a cheap bogeyman by people—often conservative and eager to enforce traditional gender roles—who want to claim there's a war on the very idea of men. These people conflate specific criticisms of a toxic streak in mainstream masculinity with universal criticisms of men. The problem with toxic masculinity has, in my view, very little to do with the core of masculinity.

I am someone who has been a boxer, a wrestler, and a student of Brazilian jiu-jitsu. Once upon a time, I could bench-press 300 lbs. I own and shoot firearms. I have been in my share of fights (enough to know I am not a tough guy) and, like my dad, have had my nose broken on multiple occasions. My skin carries plenty of scars and tattoos. I once, barehanded, fought off a hitchhiker-turned-carjacker who was armed with a knife, leaving him bloodied on a country roadside, and I have told that story over a beer more than once.

Is any of that toxic masculinity?

I don't think so.

I think strength is a virtue. I think understanding violence and self-defense is both reasonable and useful—often vital in fact. I think toughness and resilience are worthy attributes and worth celebrating.

Gender, like all components of identity, is personal, fluid, and complex. I absolutely do not think these sorts of strength are the only ways to define or express masculinity, but I am happy to include them in the conversation as well as my own, personal, evolving conceptualization.

My understanding of toxic masculinity centers on all the things that get paired with masculinity that are neither strengths nor assets to an individual or a community, things that, in fact, harm and diminish men.

Stepping on a mouse in the woods has no practical value. It's not strong. It's not brave. It neither feeds nor protects anyone.

It is cruelty for cruelty's sake.

It is lashing out at the weak as an imitation of strength.

If anything, it's about cultivating a general disregard for life.

That's the heart of toxic masculinity. A twisted worship of a false image of strength that makes any perceived weakness or deviation from the norm into a devil. It demonizes physical weakness, yes, but also emotional weakness, often interpreted as sensitivity, empathy, or even complex or nuanced thinking. This was part of the catalyst for my own shame and avoidance in addressing my mental illness. In a very real sense, I was ready to die before admitting I needed help and spent many years suffering in silence because I bought into the idea that silence was tougher than honesty and vulnerability when, in reality, the opposite is true.

I can use my time at Relson Gracie's Brazilian jiu-jitsu (BJJ) academy in Westerville, Ohio, as an example. I was neither a talented nor particularly dedicated student of grappling, not compared to my teammates, but I attended for several years and earned my blue belt. A blue belt is only the second belt rank of five in BJJ, but only about 10 percent of new students earn it.

In my time training, I saw men arrive who were clearly invested in a tough guy persona.

They would walk through the door, steely-eyed and grim, take their first class and then lose again and again and again in sparring with smaller but more experienced students. They would be shown their vulnerability. They would come face-to-face with the truth that ego is not strength, that self-image doesn't win fights, that acting hard is not actually being hard. And, in many cases, they just couldn't take it. These tough guys would leave looking shaken or enraged and not return to a second class, essentially deciding that it was better to feel tough than to be tough. It's always easier to build artifice than capability.

This is a well-known phenomenon in jiu-jitsu. Ego standing in the way of progress. I've witnessed parallel situations in skills ranging from weightlifting to creative writing. Men who are taught that acting strong and skilled is what matters will often struggle to let go of that pretense in order to actually practice and develop.

Ego is, like the fear that drove me to procrastinate on papers and hide my true self from scrutiny, another way to reject earnestness and retreat from vulnerability.

One problem with the empty, blowhard ego is that it has no tolerance for failure—equating it with weakness—but failure is the only way to improve.

The issue with toxic masculinity is not that it's out of fashion or arbitrarily maligned. The problem is that it's toxic to men. It makes us smaller than we are. It boxes us in, conjuring illusions in place of reality then trapping us within them. Of course, it's not only toxic to men. It's toxic to everyone. Thoughtless ego and mistaking arrogant cruelty for strength damages relationships, families, and communities.

I can sympathize with wounded egos, and I am well-acquainted with the fear of vulnerability.

I won't pretend that I ever enjoy failure or am always wise enough to see it as a path to progress.

I went into my first jiu-jitsu class with experience in wrestling and boxing. Shotokan karate and Shorin-ryu karate. I weighed 230 lbs. and was generally considered stronger than I looked (as a guy who looked pretty strong).

After class, a fifteen-year-old named Tyler effortlessly beat me in sparring again and again with the kind of efficient gentleness that made me certain he was being very friendly with me as he forced me to tap-out in a variety of ways I had only seen in televised MMA fights. He looked thoughtful and maybe a little bored as he did it. Meanwhile, I almost puked from exertion and had to wait for my dizziness to ease before driving home. I was in my thirties and probably outweighed him by 70 lbs.

It can be startling to learn exactly how tough or how smart you are not.

It can crack your foundations.

Yet, with the right attitude, a stubborn, adaptable, shovel-cracking, forsythia attitude, this can be exciting rather than crushing. It means that the potential for growth and skill that lies ahead was hitherto unimagined. It means that there is so much more to know.

Again and again, in martial arts and strength training, among other skillsets, I have found that thoughtful, open, generous men who know how to learn from defeat are the ones who progress and become truly formidable. Earnestness and acceptance of vulnerability are not weaknesses, but certain fragile models of masculinity would have us believe that no vulnerability is acceptable, even actively constructive vulnerability.

Toxic masculinity teaches men to avoid learning so they never feel the vulnerability of being a beginner.

The kind of toxic masculinity I grew up with also encouraged a broad rejection of emotional investment in just about anything. In the culture that raised me, empathy was a skill for women and we all wanted ways to distance ourselves from such feminine attributes.

Honestly, it makes sense. I am not surprised that the rejection of empathy was a more prominent component of masculinity than any celebration of real strength. It is, as ever, an easier, lazier path to define identity by what we are not rather than by articulating and demon-strating what we are. From bigotry to thoughtless political tribalism, the "us-versus-them" shortcut to identity remains exceedingly popular for a reason.

"I hate that person because they aren't on my team" will always be an easier touchstone than "Here is my philosophy and I stand behind it."

Toxic masculinity craves conformity and orthodoxy because it lacks courage. It isn't confident it can survive in a world where other kinds of meaning also have merit. Your truth is yours, not an unshakable reality for all people everywhere. Coming to terms with that basic fact takes strength. It takes strength to recognize that your truth is a built-thing and still have the courage and will to affirm it.

It takes strength to recognize that the mere existence of other people holding different truths from your own is not a threat but a simple inevitability on a planet with billions of minds, perspectives, and experiences. The giant, diverse family of humanity is not obligated to reflect your worldview simply because you lack the fortitude to recognize that concepts of subjective meaning, concepts like masculinity, are inherently personal.

Our personal meaning matters because we make it matter.

That's the magic.

Others cannot be expected to do this work for us.

It's easier to pretend that your meaning is just a reflection of the real world.

Easier and willfully dishonest.

Toxic masculinity is toxic, but it is the alluring toxicity of an opiate, a way of avoiding strenuous or unpleasant realities. In this way, it's doubly dangerous.

❖

I once showed my parents a draft of an essay I'd written about toxic masculinity and never published. The essay included the fieldmouse story and many of the points I have made here.

My mom had a conversation with my dad about it and wrote a thoughtful email in response.

Mom wrote, "Dad's and my conversation after reading it was about the past generations of men (his father, my father, our grandfathers), and how 'toxic' masculinity was a necessary skill for survival. Both of our fathers went to war—mine even killed other men. They could only teach their sons what they knew and what they had been taught—and what 'worked' for them in their lives."

I never met my mom's dad, but I knew Grandpa Kelley was an Army machine gunner and a cook in World War II.

I once talked to my paternal grandfather about his time in the Navy for a report I wrote in school.

He told me a story about his time burying countless rotting bodies somewhere in the Pacific following in the wake of island battles, how limbs would detach from corpses as he tried to drag them toward graves being dug by excavators and bulldozers.

Grandpa looked sick as he told me.

When I told my dad about it, he said he had never heard that story before.

I know a man who returned from war and talked about how staying aggressive was part of staying safe.

That made sense to me, in the context of a warzone.

But are wartime lessons good reasons to teach numbness and meanness as cultural values to half our population? If a public figure is an asshole, should we call him tough or strong or a model for masculinity because we fear one day our sons may need to stay aggressive on a battlefield, or may need to shut off their emotions while burying bodies?

Warfare has long been linked to masculinity.

Once, perhaps, the key attributes of warfare were physical strength and resilience, but in a world of nuclear weapons and robotics, is this still the case? How much of a nation's lethality in modern warfare is based on detachment from empathy? On numbness? On a rejection of human feeling?

I grew up in a household that contained far more guns than people. Many of my peers did as well. Most of us carried knives and many of us had experience hunting. In such an environment, realistically, the power to end life is mostly based on a willingness to do it. Couple that with a culture that teaches men that disrespect should be met with force, a culture that says vulnerability and weakness are intolerable, a culture that says, "Stay aggressive," and it creates a world for men in which power is worshiped, and power is, in the context of modern tools of violence, largely founded on numbness.

Numbness won't win a fistfight, but it can easily put a bullet through someone. I don't hear news stories about mass fistfights. I hear news stories about mass shootings.

I was a high school senior when the Columbine High School massacre occurred.

At the time, I remember choosing to keep quiet when I heard so many adults, politicians, and media personalities say they couldn't believe it happened, that they just couldn't imagine what could prompt such a tragedy.

I could believe it.

I could imagine it.

It made perfect sense to me.

The stranger idea for me was that so many adults were surprised.

I remember wondering if they were pretending.

I still wonder.

These days, it feels like there's a mass shooting every week in my country.

Now, more than ever, the surprise seems performative.

In the context of toxic masculinity, disrespect can be anything from exposing vulnerability to threatening or denying one's concept of objective Truth—and disrespect should be met with violence. In a nation where I can buy a gun at Walmart, the real path to efficacy in violence is linked with the power to turn off emotion and empathy.

Killing is an unambiguous path to masculine virtue, celebrated in our media and our family stories.

So, what part of these shootings is a surprise?

Cheap and vicious. Proudly, stubbornly unfeeling. Toxic masculinity is a concrete barrier standing in the way of real emotional connection, standing in the way of mental health care, standing in the way of alternate paths to self-expression and constructive ways of processing anger and sadness. Meanwhile, we have endless encouragement toward violence and easy access to the tools of killing.

Certainty is easier than inquiry.

Feeling tough is easier than being tough.

Loud gets more attention than thoughtful.

Our culture defends the takers and the blowhards while it mocks, devalues, and marginalizes the thinkers and the feelers, the caregivers and the teachers.

Consider school funding.

Showing the practicality of funding STEM programs in school is easier than illustrating the more abstract value of arts and humanities. After all, *will empathy and creativity get you a job*? And, *just between you and me, aren't those second-tier, feminine skillsets anyway*? I can't tell you the number of times I told somebody I wanted to be an English major, and they joked about me working in fast food the rest of my life.

These are the same blunt, narrow, inhuman cultural values that had me believing for decades that mental illness was a shame worth hiding, whatever the cost.

There's a reason nearly all mass shooters are men, and it isn't because women are never furious or disrespected.

It's because men are taught that one of our chief virtues is the ability to ignore the value of other lives and women are, broadly speaking, taught the opposite.

I think of TV pundits who wonder how local teachers and parents could have missed all the warning signs surrounding any of these mass-murdering young men. When I was a teenager, I had a pegboard wall in my room to display all my knives and weapons. I collected books on lethal combat techniques, poisons, and making homemade explosives. I don't recall any of these facts ringing alarm bells in any real, tangible way. I look back and see a young version of me that was odd and private and dangerous and just not that unusual among my peer group. I had plenty of access to firearms. I struggled with mental illness in a culture that discouraged me from discussing (or having) feelings beyond anger, lust, and ambition. My teen years tell me that many of the warning signs that might indicate an impending tragedy aren't rare enough to function as red flags, not in a culture that sows thickets of red flags in the minds of young men, conflating cruelty and viciousness with strength and honor while holding up capacity for violence as one of the very few objective merits men can attain. I was raised among parades celebrating combat veterans. Not teachers. Not scientists. Not caregivers. Not poets. It shouldn't be a struggle to understand how our young men turn to guns and rage and hometown warfare. We teach them that men are ready for war. We give them little else.

To be clear, of course I am intellectually disturbed by my lack of shock at mass shootings. I feel certain that reports of mass murder by young men should fill me with horrible surprise and confusion. I also feel certain that not feeling such things, for a span of over twenty years of regular, high-profile mass murders, means that something in me and in my culture is deeply, profoundly broken and we can't fix it until we start talking about it honestly.

In the small-town Midwest of my youth, I didn't have tools to manage my mental illness or my anger or my endless anxiety at being a square peg in the round hole of traditional public school, but I did have the tools to kill.

Yes, I'm over generalizing based on my own upbringing and experiences.

No, I don't think I have reached a point where I am somehow outside or above these patterns because I am uniquely wise or insightful, but I do think some of my experiences have given me alternate paths to understanding.

To state it crassly, I think that getting my ass kicked in activities like boxing or jiu-jitsu complicates and deepens my understanding of toughness.

I think getting my ass kicked by mental illness and encountering diverse voices and perspectives complicates and deepens my understanding of empathy and society.

I think getting my ass kicked by confronting the science of climate change and by interrogating our cultural tendency to commodify and dismiss the living world complicates and deepens my understanding of nature and our connection to it.

I don't have all the answers nor all the questions.

Yet one piece of the puzzle, one integral flaw in our system and our conception of masculinity in particular, arises from the fact that we don't teach men the value and honor in getting their asses kicked (physically, intellectually, or emotionally), in losing in order to grow, in dropping the cheap, easy mask of toughness in order to be something more than a mask, in being brave enough to welcome physical and emotional vulnerability and be changed by it.

We don't teach the value in sitting with the deep emotions, the hurt, the confusion, nor the value in accepting that those emotional punches will and should send us sprawling, should crack our jaws and leave us changed, leave us stunned and reawakening inside a new perspective, our cheeks resting against the cold pavement.

We do not teach men that one of the powers that matters most is the power to craft and honor our own truths so that we don't rely so heavily on others' perceptions and prescriptions for our meaning, masculinity, validation, and self-worth.

Understanding violence and self-defense are worthwhile endeavors.

There are some threats and injustices that are only addressed through violence.

That doesn't mean the lion's share of our pop culture should glorify men killing or beating people, and it doesn't make being an ass on social media the same as being aggressive in a combat situation.

My dad says he remembers the mouse-stomping conversation and that he was probably thinking about keeping the rodent out of the house. We had a mouse problem. Most every rural home has a mouse problem.

Sure, we can kill mice even before they enter our homes.

It's true.

And in this age of technology and weaponry, we could kill literally everything before it has a chance to do anything.

I'm an American. I suppose the United States could preemptively use nuclear weapons against many other nations before they have the chance or the strength to be a real threat.

Think of the weight that places on our philosophy, on our meaning.

On a personal scale, we could shoot our neighbors or our classmates, a simple pull of the trigger, before they have a chance to hurt our feelings again or make us question our sense of self.

We can all think of examples of men who made that choice.

So, what happens to us, and to the world which sustains us, when, given the state of weaponry and our access to it, we equate a cold lack of restraint with toughness and with masculinity?

What is the cost of such a philosophy in a world in which humans hold the technology to kill every living thing on the planet many times over? In a world in which so much of modern masculinity rests on coupling toughness with honor and honor with violence and violence with numbness? A world where we gut arts programs, sending one more signal to young people that they are here to be machines of commerce and productivity? When I was a weird, angry teenager, I had poetry and drama and art alongside martial arts and other sports. If I hadn't had those outlets, I'm not sure how I would have handled it. I know my thoughts tended to bend toward violence.

I'm certainly guilty of aggrandizing violence.

I get it.

It's fun.

Yes, I am a thoughtful nature poet.

Also, I find violent video games to be profoundly relaxing and cathartic. It's simple and satisfying to have the unambiguous task of destroying a fantasy bandit or a rampaging cannibal mutant.

These games drop me into settings which are big and mean and my every brutal triumph over death feels like a real victory, especially when my character started as an underdog. It's the fantasy of having both lethal power and an unambiguously righteous reason to use it, a classic masculine power fantasy.

But in the real world, humanity is not the underdog, and our collective strength has very little to do with anyone's individual power, bravery, muscle mass, or emotional numbness.

It's seductive and comforting to imagine we live in a vast, limitless world in which we can express the full extent of our power (and violence), because we are small and our planet is large.

It's comforting to think that our fathers will never grow old, and that the institutions and cultural benchmarks we know, heavy with the inertia of history, are constant and unshakable.

Comforting and false.

The truth is, life is fleeting and fragile, and the world will not stop us from destroying it. Cruelty and numbness aren't the kind of power our survival needs now. We need empathy and nuance. We need understanding and community.

No real strength is solitary. Not in an individual or social sense. Certainly not in an ecological sense. Communal strength needs communal values, and that means privileging empathy.

I expect there are people out there who do indeed think destroying the world through violence or through dominating and consuming nature into a noxious husk is a kind of power. If that is power, it's power with an ugly, silent aftermath. It's a power antithetical to life and joy.

Too many who worship violence, like those who worship money, start to think of it as a good in and of itself, divorced from other purpose.

As someone who values the skills of fighting and shooting, I often think of them in terms of preserving life, of protecting others. There can certainly be truth to that, but on a philosophical level, I always arrive at the conclusion that violence is a lesser, B-grade kind of power.

The power to destroy is only the power to hasten endings that would have arrived in their own time with or without our influence. Destruction is in service to the inevitable, in service to a master that, in the fullness of time, has no real use for us.

That's not power.

Care is power. Building and nurturing is power. Creativity and fostering growth are power. Teaching the young and ourselves through welcoming courageous vulnerability is power.

Sometimes there are moral reasons for violence. But it is toxic nonsense that so many of our young men think of violence as their only real pathway to power and, in some cases, their only real function and purpose in this world.

I sometimes wonder if part of the issue is civilization itself. The comfort and structure of it. The isolation and disconnection of a world in which many of us have little real need to connect with our communities in a broad, survival sense. In a comfortable enough context, being tough can seem like a fashionable affectation, and being kind can seem like a purposeless nuisance.

I have found clarity in stepping outside these contexts.

For example, I enjoy walking through the woods at night.

Weird, I know. Are you surprised?

When I take a light, I will often see fieldmice dart across the path, so fast I can barely register their presence, almost an optical illusion.

Yet, more often, I am quite content to walk in complete darkness.

It's rare, but on more than one occasion, I have met another man walking the opposite way on his own night hike.

It's an odd, electric sort of feeling, meeting an unexpected stranger in the middle of a dark nowhere.

There is always an "Are you friendly?" sort of tension in the air when you find yourself in close proximity to a stranger, but it is magnified a hundredfold when the meeting takes place in the lightless woods.

Do you put a hand on a weapon?

Do you raise an open palm?

From how far away do you offer a greeting, to seem friendly but not overeager?

I have seen men exchange fighting words in a bar when they were certain their friends wouldn't actually let them fight.

I have seen men exchange fighting words online where fighting is safely off the table.

I have seen dogs yapping through a fence when I very much doubted they would actually want to fight if the boundary was removed.

When I have met these strangers in the darkness, more often than not, a heightened civility arises.

Big greetings. Big smiles. Warm tones.

"Hey, man. Nice night. How's it going?"

"How ya doin' brother?"

I've been asked for a lighter. I've been asked for directions. I've been asked if I heard that great horned owl call a while back.

These interactions puzzled me at first, the way they felt both different and unusually companionable.

Then, I got it.

There is no default civilization present in these meetings.

The lack of obvious signs of civilization also meant there was a conspicuous absence of masculine posturing. There was no chain link fence through which to bark. There was no audience for whom to perform toughness. There was nobody to see us, stop us, or punish us if we decided to kill one another.

Just us and the dark trees.

Far from help and community.

The only civilization these strangers and I had in the moment of our meeting was what we intentionally created ourselves. In those moments, the whole of human civilization exists as a collaboration between two men in the dark woods. In all that wild, inhuman expanse, there is an instinctual draw to connect and build something human together, to fall back on the core strength of the human animal—community.

Those experiences lay bare an essential fact about our civilization and our culture that can be easy to overlook.

They're a choice.

They're a mutual construct.

There is something startling about encountering a reminder of this fact.

It's an odd, vulnerable circumstance to meet an unexpected stranger on a dark path.

It's a reminder of just how safe we are not.

It's a reminder that human culture and civilization are not a part of the natural landscape; they are, like roads and electric lights, technologies.

They are built things, another kind of crafted meaning.

Civilization is a candle held in the darkness.

When enough of us hold the candles, it can seem very much like daylight, but the night has not actually retreated. The illusion of modernity is that we are no longer holding the candles, that they are self-holding and the night itself is an irrelevant relic of a bygone era. It isn't so. Even now, we hold those candles together, by choice.

These encounters make me hopeful. In reminding me that civilization and culture are constructs, they remind me that both of those concepts are shaped by human minds and choices.

So, they can and will change.

Perhaps the toughest, strongest aspect of the human animal is our ability to change, both ourselves and our environments, to adapt ourselves and our technologies to fit our current needs.

Toxic masculinity, as a cultural technology, is not meeting our current needs.

With enough thought and attention, we can leave it behind and create something better.

The choice is very much ours to make.

❖

There's a deep, bone-colored utility sink in my basement that I rarely use.

That July, I visited it to rinse spilled laundry detergent from my fingers and found two withered fieldmouse corpses nestled together.

Dry and featherlight, they must have been there since the winter, when mice become especially dedicated to finding a way indoors.

They had fallen in, hidden themselves away as best they could in the shallow recess of the drain, and died of dehydration.

I saw them, felt the circumstances of their deaths unfurl in my mind like a morning glory opening to the dawn, and my heart rose into my throat.

I couldn't imagine how they fell in, but I heard the hollow thunk of their little bodies hitting plastic.

I heard the frantic skritching of tiny claws that couldn't find purchase on the smooth, vertical walls of the sink.

I thought of being trapped in an implacable, shelterless canyon of unnatural whiteness.

It made me want to cry, and I shook my head at the impulse.

My basement is a dim place of cobwebs, bare lightbulbs, and exposed floor joists toothy with the ends of nails. I stood there, looking at the dead mice, feeling ambushed by a tide of grief—startled by the intensity of the emotion.

After all, I kill mice with snap-traps.

I have intentionally engineered the death of many mice.

I don't do it out of malice. I do it because sharing my space with them could expose my family to hantavirus, a rare but potentially lethal disease spread by rodents, and neither live traps nor deterrents like peppermint oil have proved effective.

In theory, I should have adjusted to the idea of killing mice.

But the death in the sink was different and I wasn't inoculated to the grief of it.

Even though it would be easier to do so, I am not willing to let myself think of a mouse as an inanimate object. Because of empathy and imagination, I sensed the difference between a quick snap of finality and a slow death in an unfathomable oubliette, exposed and starving in an inescapable blank box, scared and isolated.

I removed the tiny bodies and draped an old blanket into the sink and down to the chipped concrete floor, creating an escape route.

I started to scold myself for such an impractical, sentimental act. Allowing mice to escape from the sink increased the risk of finding droppings on the kitchen counter and chewed holes in the pretzel bag, but I knew I couldn't check the sink every day, and I also couldn't tolerate letting any more fellow creatures suffer and exit life in such a forgotten, forsaken way.

My inner critic rolled his eyes at me, questioning my sense, logic, and yes, my masculinity. Mercy is not masculine. Sentimentality is not masculine. I should remove the blanket and be a man about it. Mice are vermin.

There was a time when I wouldn't have consciously noticed such thoughts, I would have just bowed to the cultural mandate to be unfeeling, removed the blanket and pretended not to feel the pain of a nameless wound opening on my spirit, but I was becoming more aware of the mechanisms of my own mind.

These choices matter.

Distinctions of feeling matter.

Imaginative empathy and sentimentality matter.

The version of masculinity that I call toxic is the version that says such emotional distinctions are useless, that a dead mouse is a dead mouse is a dead mouse. Utility is utility. Cold, efficient logic is a masculine virtue and everything else is a plaything or a distraction. And if you practice such thinking, it will certainly become true.

That's how meaning works. That's the dark side of the magic.

I often talk about the importance of making meaning, and there is a corollary that I tend to avoid saying, at least with these exact words, because it is often misinterpreted—the universe is a meaningless place.

Friend, this is not a bleak thought.

I'm not being grim or nihilistic.

I'm recognizing that the kind of meaning that matters to humans is necessarily a human invention. For me, saying there is no meaning in the universe is like saying the universe does not produce its own top hats or toasters. Of course it doesn't. Why would it? Those are human creations made for human purposes. What would they be doing here without us?

The idea that a meaningless universe is cruel to humans is like thinking the world is cruel to honeybees for not having readymade honey on hand in preparation for their arrival.

The universe doesn't need honey.

Bees do.

The universe doesn't need human meaning.

We do.

This is one more vital problem with the impulse to shut down feeling and empathy, to adopt toxic masculinity as a facet of our identities.

If men decide that cruelty is what matters, then, in a subjective sense, they will be correct.

And if cruelty is correct, it will follow that the justification for this is also correct, that the self is what matters most of all. Cruelty is a rejection of empathy, a rejection of the value and dignity of other creatures.

If a man decides his version of meaning is a kind of solipsistic faux logic that mandates self-interest as king, that only his wants and needs matter, then that is certainly the meaning he will have as a guiding light.

We all know people like this.

Certain CEOs and political leaders come to mind.

Likely, you know folks like this a lot closer to home.

It can be a functional kind of personal meaning to adopt, at least in terms of social and financial gain. Indeed, our current culture is largely built to reward this personal philosophy. Take what you can. Trample anyone in your way.

Yes, any one of us may choose to view all mice as vermin.

Men can choose the same outlook for women or anything we deem feminine.

We can view other cultures and other viewpoints as vermin too. Of course we can. Meaning is ours to construct. If enough of us share a meaning, it becomes a culture that reenforces itself.

Human minds are powerful. The awe-inspiring truth is that as impermanent beings in an impermanent universe, nothing is important until we imbue it with importance. We can dismiss the importance of whatever we like. Feelings. Nature. Culture. Art. The long-term survival of Earth's ecosystems.

We come up with all sorts of reasons for such dismissals. Profit. Logic. Practicality. Manifest Destiny. Traditional masculinity. Readiness for wartime.

There's a kind of comfort in this, a kind of austere simplicity that makes life seem pleasantly straightforward.

Self-interest is logical and logic is our best guide. This works well if you don't think about it very hard. If you ignore that pure self-interest as a member of a social species living in close proximity ignores the broader context of our survival and happiness.

Unchecked self-interest as a human, a creature whose most potent strengths rest on communal endeavor and decentralized expertise, is not logical.

Pure self-interest in a world of linked organisms surviving through complex, balanced relationships of mutual dependence, relationships we do not fully understand and certainly do not control, is not logical.

Cruelty. Arrogance. Worship of self-interest over community. Dismissal of empathy, emotion, and nuance. Commodification of nature and people. These are the defining characteristics of toxic masculinity, and our health and future demands that we actively notice and reject these corrosive values.

Defenders of the tenets of toxic masculinity pretend that critique of such principles are an attack on all men or all masculinity. They aren't. They're a critique of the toxic part.

Part of the draw to enforce a strict, zealous orthodoxy of masculinity arises from a truth many men are uncomfortable confronting: there just isn't that big a difference between men and women.

I understand that gender matters and is meaningful. It would be disingenuous and dismissive to claim it isn't. That said, it was a startling epiphany in my young life to realize that gender is not as meaningful as I was raised to believe.

As a young man, I don't think I thought of women as people. I mean, I knew they were human, just not human in the same way as me.

That was a hard sentence to write and a hard truth to admit.

But it's true.

The conversations I heard around men and women were always focused on fundamental differences. I had internalized the viewpoint that women contained an odd, pervasive otherness. Separate and mysterious. I remember in high school wanting a girlfriend, but feeling deep anxiety about the utter unknowability of the opposite sex.

My relationships and prospects for romance changed for the better when I realized, *Oh, women are like me.* They have wants and dreams. Some of them have similar interests to mine. They also want love and connection.

We are all far more alike than different.

There's something truly absurd about a culture in which I had to realize that 51 percent of my species are not inherently different creatures.

I like to hope we will do better for the next generation of boys.

There is nothing toxic about men.

There is a great deal of toxicity in teaching men to be mean, thoughtless takers and users.

There is a great deal of toxicity and cowardice in teaching men to be suspicious of emotion and vulnerability.

There is a great deal of toxicity and danger in teaching boys that their best path to honor is violence.

There is a great deal of toxicity and injustice in a culture that won't hold men accountable for the damage arising from such teachings.

There is a great deal of toxicity and alienation in teaching boys that women are distinct and lesser creatures.

The cowardice, cruelty, arrogance, and ego that define toxic masculinity are antithetical to actual strength and, in my worldview, actual masculinity. Real strength is thoughtful. Real power rejects cheap conformity and empty, performative cruelty. Men, past and present, are not blunt, unfeeling tools. We are not hammers and the world is not a nail.

This matters.

Uncovering and confronting the influence of toxic masculinity in my own mind, separating it from my personal understanding of my own masculinity, has been necessary in making progress toward happiness, connection, and mental health.

My own toxic masculinity was just one more way my depression could defend itself.

Toxic masculinity didn't empower me.

It didn't make me tougher or stronger or smarter. It made me smaller. It made me less able to adapt, to face hardship. It trapped me within the expectations of cowardly, unimaginative, superficial men.

A broader, cultural conversation about toxic masculinity will be vital in our efforts to preserve our planet and to build a better, happier society for all people.

Physical strength and toughness are virtues. Courage is a virtue. Resilience and determination are virtues. Sometimes, violence is necessary, and it is a skill worth developing.

Yet, when I must kill a mouse, I want to feel the emotional pain of it. I want to choose to feel it.

I want that gut punch to take my breath.

Because, in the long term, not feeling it is much worse. I write this as a man whose pain has washed away years of memory. I write this as a man who has had to gather the splinters of his shattered wonder and emotional wellbeing and slowly, intentionally build it anew.

Choosing not to feel because hurt lurks in the world is akin to never risking our hearts in love or never walking the woods for fear of unexpected storms.

Worshiping power or money or strength or cruel detachment will not make us safe.

We are not safe.

We will never be safe.

That's why we need courage, to become something we find worthy and admirable while living a limited, temporary life. We need courage to help others in this same journey.

Living, loving, learning, nurturing, and empathizing are brave acts that, when done thoughtfully and honestly, require intentional acceptance of vulnerability. For people like me who were raised to value both ironic detachment and stoic, masculine detachment, that sentence feels threatening, trite, and cloying. It has taken much of my life to realize that the larger passive copout and the larger threat to self is refusing to engage in earnest emotion.

For those who believe self-interest is still their surest guidepost, I offer a question.

What do we give-up when we adopt the numb, self-obsessed ego and cruel disdain of toxic masculinity as our key cultural values?

What does it cost us to exclude the worth of other creatures in order to prioritize our own limitless greed and hunger in all things?

We give up wonder.

We give up our best pathways to real personal growth and strength.

We give up diversity of thought, interest, and skill.

We depopulate the world of other worthy beings and minds, alchemizing our grand, complex, varied planet into a featureless box.

We become a mouse trapped in a sink.

If mice are only vermin and other lives are reduced to things for our use or pleasure, then our world is an ugly, transactional place. A place of things.

If, on the other hand, nature is our family and even fieldmice deserve our consideration, our respect and empathy, then we may live in a world worthy of our love, wonder, and curiosity. If we allow our world to be populated by wonders and make peace with our own limits and impermanence, experiencing becomes more important than owning and consuming.

The meaning we make defines the character of our world and lives. The choice to create meaning with love and gratitude is not simply a gift to the reality we shape with our ideas and actions, it is also a kindness granted to ourselves. In the end, empathy serves and honors the self far more fully than any celebration of ego.

We all suffer.

We all die.

Everything ends.

These things don't require our choice, consent, or participation.

On the other hand, wonder, love, and empathy don't simply happen on their own. We must bring them into being. That is power. That is how we exercise the strength to shape our reality. That is how we build meaning and culture that serves us now and in the future. This is the prime vocation of men, of all people, the way we can shape the world with our physical, mental, and emotional muscle.

If strength matters, this is why.

If masculinity matters, this is why.

The strength to care and learn. The strength to feel and grow. The strength to build and preserve. The strength to resist seductive ego, numbness, and arrogance.

This is what matters.

Chapter 13

LIGHTNING BUGS

SUMMER MOVED ALONG and I felt the ache of healing.

Healing, of course, is a good thing.

Growth is a good thing.

But good things aren't always comfortable.

Discovering new paths to mental health was my goal, and that summer I felt unusually confident I was achieving it.

Yet any new topography on the map of your mind can be disorienting. Even an unambiguous benefit like having more energy can be a challenge when it's new.

That sudden extra potential can ask you endless, uncomfortable questions about "what now" and "what next."

I often did not have the answers.

One evening in June, I stood at the back fence where a forsythia once grew, watching the lightning bugs bob and blink above the dusk-dimmed gravestones with the slow grace of an underwater ballet.

Leslie walked up beside me and wrapped me in a hug.

"Do you remember the first time we caught fireflies together?" she asked.

I did, but I asked her to remind me anyway.

I like that story.

Shortly after we started dating, Leslie brought me to visit her parents and her hometown in Michigan. She was born in Detroit, but grew up in Brighton, a commuter community from which her dad drove to Detroit each morning and her mom to Flint. The place felt pretty fancy to me; its little downtown was that kind of strenuous, meticulous cute that tends to mean money.

Leslie often reminds me of that trip.

Not because of the town or her parents, but because of a twilight walk we took past a fountain that local kids liked to fill with bubbles as a prank, near a scraggly field by her old high school. That evening, the place was full of lightning bugs.

Leslie grew up saying *fireflies*. I grew up saying *lightning bugs*. Now, we use both interchangeably.

We paused by the field and watched the swooping lights.

"I love fireflies," she said, "but I've never been very good at catching them."

"Oh? Would you like to hold a lightning bug?" I asked. "I don't mean to brag, but I am *very* good at catching them. I can show you how."

She smiled and took me up on my offer.

I grinned like a Jack-o-lantern and marched into the weeds.

I showed her what worked for me, creating a flat platform with my palm and gently raising it up beneath the flying insects.

These were big dipper fireflies, the most common species in North America, named for their diving, dipping flight pattern. I found that they tended to flee downward, so raising a hand up beneath them intercepted their evasive maneuver. They would alight on my open palm, turn, walk to the edge of my hand, bloom suddenly into a tiny flower of black wings, and fly soundlessly back into the gloaming.

They were being especially cooperative with me that evening, and I urged several from my hand onto Leslie's. Soon, she got the knack and was catching her own.

I suspected she already knew what she was doing, but she enjoyed my enthusiasm.

We stayed in that field until it was full dark, the little creatures shining and swooping around us, fleeting stars in fleeting constellations, producing light through a whimsically named class of enzymes called luciferase, apt for tiny, blazing, falling angels.

It wasn't until many years later that Leslie told me she thought of that evening as among the most romantic moments of her life.

"Really?" I asked. "Why romantic?"

"Because I think of my childhood home and my high school as frighteningly serious places. Catholic school. A conservative town. Places built around expectations, around guilt and anxiety. And that night remade those places into something completely different like it was the easiest thing in the world. I guess it was the whimsy of it all. Easy and transformative."

Leslie is a poet and talks like one.

I like the idea that I can't predict when I'm being romantic.

I like being a character in a loved one's history and seeing myself through that lens for a moment.

I like Leslie's insight in landing upon the word *whimsy*. I suppose a nearly thirty-year-old man grinning and concentrating on catching lightning bugs in a twilight field in Michigan qualifies as whimsical.

Whimsy often carries the connotation of being pointless, yet Leslie counts that evening among her all-time favorite memories.

There is power in whimsy. It is a cousin to magic. While my sense of magic tends to involve meaning crafted with intention and purpose, whimsy is a bit less focused, though not less important for fostering a friendship with life and a joyful, welcoming worldview. Whimsy is enthusiasm that refuses to defend itself with practicality. It is play. It is romance and nostalgia. It is finding meaning in pure feeling, respecting emotionality without tethering it to logic or narrowly defined utility. It is the practice of honoring delight for delight's sake, love for love's sake. I suppose that could be considered frivolous in the sense that whimsy isn't aimed at money or practicality, but that strikes me as a crass, harmful conception of frivolity. Is happiness really so useless?

The truth is, I need whimsy to live. I suspect many of us do.

I need to wander through insectile constellations in fields beneath the dull, copper glow of a summer sunset.

I need to chat with maple trees and share meaningful moments with herons.

I need to see myself as a dark, warm wilderness through the perspective of my gut bacteria, or an inscrutable, striding giant through the senses of a jumping spider on my kitchen counter. These are whimsical imaginings, but important all the same, important because perspective informs our reality and our choices inform our perspective. Choosing fun and wonder isn't an impractical, passive escape; it is a valuable, active method for enriching our days, an intentional way to invest ourselves in the premise that life is rich and worth our attention and effort.

I need to zoom in to the swirl of my fingerprints and see them as living maps of galaxies, or zoom out and see the water in my body as part of a continuous, abstract river flowing from storms over a blooming meadowland down to a midnight seabed where giant spider crabs pace the darkness beneath darting flashes like living lightning, bioluminescence from creatures they do not name while, high above, whale-shadows drift across the liquid sky like passing clouds.

When someone loves you, I think they can sense when you are inviting them to come inside and participate in a vital part of who you are. Leslie knew she was seeing to the core of me when I was chasing lightning bugs. It is constructive vulnerability. Romance directed inward and outward.

There is a dim, gray cavern at the center of me that is neither lit nor warmed by practicality, but whimsy can build a bright hearth there and make it into a home, both for myself and for those who would enter my life.

It is the furthest thing from frivolous.

❖

That summer evening, and many before and since, Leslie and I decided to walk in the cemetery.

When the summer moon shares the sky with a western blush, the cemetery beyond our fence is full of fluttering bats and drifting fireflies like cold embers from an absent fire. The bats don't eat the lightning bugs. And neither seem concerned with a quiet couple walking hand-in-hand.

I once told a friend that we love our evening walks among the headstones and she said it sounded "spooky."

It honestly never occurred to me to think of the cemetery as spooky, partly because of the thousand miniature gardens and grave decorations honoring the departed, partly because living in close proximity means I see the conspicuous purpose of the place from my windows every day.

A cemetery is not a Halloween decoration.

It's not a horror movie set, nor is it simply a utilitarian place to deposit our dead.

Our cemetery is an arboretum.

It is a quiet, thoughtful place in a loud world.

It's a place where I spot young Cooper's hawks loafing on statuary and bald eagles resting high in the pines on their journeys between lakes and rivers.

When school is in, local sports teams jog the looping gravel lanes between the neighborhoods of graves. I see picnics with lost family members. I see children flying kites. I see witchy teenagers feeling witchy together and history teachers leading fieldtrips.

The cemetery is also a nostalgic place for me, reminding me of my walks in the big, old cemeteries of Newark and Athens in my teen and college years. Cedar Hill was my favorite cemetery in my hometown. The West State Street cemetery was a frequent destination during my years at Ohio University. These were quiet, wooded places where my constant thoughts of mortality didn't seem so alien and out of step with the rest of life.

When I told Leslie about my friend's "spooky" comment, she wrinkled her nose and said she thinks our cemetery is "adorable."

"All cemeteries are adorable because they are huge, impractical monuments dedicated to humans loving and remembering each other. That's the whole point of them. It's sweet."

I understood what she meant.

Green space dotted with little stone love notes.

Cemeteries aren't spooky.

They're whimsical.

My office window faces north toward the back of our house. From my desk, three times this week, I've seen someone park, get out, pick leaves and weeds from a grave and then just stand or kneel with a hand on the stone.

One, I think, was praying.

One just knelt there and seemed to be chatting.

One left fresh flowers and a solar-powered LED lantern that pulses with the soft, green-gold light of a firefly until its power fades in the small hours of the morning.

Occasionally, I see these visitors rise and look around half-embarrassed to be doing something so intimate and impractical in a public space, but most seem confident in their errand.

Evergreen grave blankets in the winter.

Pops of color planted in the spring, impatiens and zinnias.

Families gather in our cemetery to watch the local fireworks on the Fourth of July; some, no doubt, watch in the company of missed loved ones.

The behavior I witness in the cemetery ranges from cheerful to somber, but the nature of the place paints all of it as sacred and contemplative.

My favorite places in the world are the ones where money seems frivolous and feeling seems practical. Meaning becomes suddenly visible against a sprawling background of life and history. Forests are one. Libraries too. And certainly cemeteries.

I don't belong to any of the religions invoked on the headstones, but it's hard to deny my own associations between burial and the sacred, a literal return to the earth.

Sowing seeds.

Entrusting treasures and memories to the ground for safekeeping.

A reunion with the mother of all the life we know.

The sacred association isn't really about the dead. It's about the living, seeing natural ritual, a special kind of meaning-making that's rooted in connection, gratitude, and love. Rooted in whimsy. Feeling for feeling's sake.

The sacred in the cemetery is alive and active, a matter of choice and effort.

People pay large sums for those stone love notes and statues. They give time and resources to make their visits, to tend their small gardens and leave their tributes. They turn away from other, perhaps more utilitarian endeavors to focus on love and remembrance.

These things are not practical in our typical understanding, yet we need them. I witness that need daily. We need them like we need to climb trees and catch lightning bugs. We need them because, sometimes, it's important to let our constructive emotions drive.

The cemetery is about love and meaning. It's also about people grappling with impermanence and employing hard stone as a barrier against it. A firm foothold in the idea of forever.

Not the reality of forever, but the idea.

In ritual, it's the idea that matters, because we make it matter.

Such is often the nature of metaphor and ritual.

Meaning and magic.

Identity and whimsy.

The oldest stones in our cemetery are already eroding back into the earth. Yet they did their jobs. They stood for concepts too big to hold in the minds of those they served. Love notes, written in stone, sent and received. Acts guided by whimsy. Meaning made tangible, for a while.

That's the point of ritual and often the point of whimsy, to give shape to something essential and shapeless. I couldn't draw a map of my feelings for Leslie, but I could catch fireflies with her. We can't measure and reconstruct the ways our lost loved ones touched our lives, but we try anyway and the effort is itself fulfilling the goal. We dedicate ritual to our remembrance. We dedicate acreage and etched marble. We work to give our feelings form and weight.

I confidently call these things essential, yet I can intuit that a majority of my culture doesn't view either catching fireflies or planting a peony in a cemetery as practical matters.

Practicality gives us bread and water—it fulfills necessities.

Except my illness has taught me that those are not our only necessities. When depression strips away your natural pleasure and whimsy, rebuilding those impulses can be an instructive process. I am not merely leaning on my subjective experiences when I argue that our necessities are far broader than our biological requirements for life. Our written, oral, and archaeological records of human art and ritual insist that the bare practicality of our physical wellbeing has never been enough for us.

If I am critical of mainstream views of practicality, it is because those views have been corrosive, destabilizing forces in my life. In many of our lives, I suspect.

As the enemy of whimsy, practicality registers as both a favorite weapon of my depression and part of a broader cultural distortion that seems to consider nature, when it is not producing obvious profits, likewise frivolous and impractical. The result? We have mass extinction, a climate crisis, and a new generation grappling with terms like "late-stage capitalism" among record levels of income inequality and profound, looming ecological and civil threats.

Practicality has proven impractical, at least as it's currently conceptualized in mainstream American culture.

We increasingly have the technology to work less, but we lack the philosophy to benefit from such advancements, to turn away from easily measurable productivity as our valuations for human life and endeavor. We're great at producing food, but we'd rather throw it away than give it to someone without the means to pay. How can we determine they deserve food if they cannot pay?

Our distorted view of practicality, and the value we place on it as a guiding principle, is broken.

It's breaking our kin.

It's breaking our world.

As for me, I wouldn't mind if certain folks wanted to spend their finite hours trying to amass more abundance than anyone else. I don't feel compelled to abolish wealth or make everyone's values mirror my own. I do not need everyone's meaning to be my meaning. That's not my business. But it becomes my business when such folks build their games of excess contingent upon the suffering of huge swaths of humanity and sweeping ecological degradation. It becomes everyone's business.

Sure, have your battles of leverage and numbers, but have it in the context of communally upheld protections for the environment we all share and a higher baseline standard of living and dignity for all peoples, divorced from productivity, founded on the idea that we are all inherently co-owners of human progress and fellow stakeholders in the march of human ingenuity, at least insomuch as it applies to collaborating with nature to produce the necessities of life and health.

I think such a model would still allow plenty of room for competition and conquest for those driven by such things.

I certainly don't think providing everyone with baseline dignity and necessities would end human striving, work, ingenuity, or ambition.

"Work or starve" is not the recipe for creativity, innovation, or productive risk-taking that some people seem to think it is.

When I talk to my peers about work, a majority of them tell me they work to avoid consequences. Unpaid bills. Loss of healthcare. Homelessness. Endless other variations on that theme.

They aren't working in pursuit of something. They are working in avoidance of something. For many, especially those without the safety nets of robust support systems or family wealth, that's the nature of our society.

In the minds of many Americans, we traded "the pursuit of happiness" for "fend off desperation" long ago. Settling for such grim motivations feels like an absolute failure of imagination for creatures who are natural crafters of meaning, for creatures who need whimsy and ritual to thrive.

I don't believe protecting the Earth while meeting our needs is an insurmountable obstacle for our will and intellect.

Can I make these things happen?

No. I am neither smart, charismatic, nor ruthless enough. Nor am I arrogant enough to think I know what is best for all humanity.

I perceive the world as flawed.

I do not control or own global outcomes.

I place my hand against the colossal stone of society and economics and press in the direction of my values.

I cannot shift it by myself, so I must hope others are pressing in the same direction.

Lately, I wonder if I can feel a hint of movement beneath my fingertips.

There is a feeling of unsustainability in our way of being that I am far from the first to notice.

We can't discuss mental health or nature completely divorced from those cultural forces that threaten our peace and sustainability. Context matters.

I know that I need whimsy, that I need to lift up joy as a right I earn just by being alive, akin to my intrinsic worth, yet that knowledge is often met by a rebuttal from our economic and cultural climate.

An invasive string of words, tough and stubborn as forsythia roots. A thought that is always hungry.

It arrives during quiet moments of contentment. A moment touching moss by a still pool. Closing my eyes to focus on a birdsong floating over the cemetery.

I'm not being productive enough.

I do my best not to hate these words.

I do my best to say, "Ah, good, you're here. I have business with you."

I do have business with them.

I must bundle these words in an old blanket and sit them in front of a fire on a crisp night. I must heal them with wordless wind and smoke.

If I cannot change the words, then I will try to change my own understanding of them.

Productivity can, after all, mean many different things.

I hope I can give these hungry words something worth savoring.

I hope I can give the same to the culture that made their hunger terrible.

Hunger, like fear, is natural.

Eat to live.

But hunger as a virtue?

Hunger as an unending, unanswerable drive?

Hunger and hustle.

Somehow our idea of productivity became insatiability.

Perhaps because when productivity stands forever in the shadow of desperation, productivity is no longer about obtaining anything. It's about escaping something, something that never vanishes because we won't allow it to vanish. We won't allow excess food to feed the hungry. We won't allow unoccupied housing to shelter the homeless.

I suppose I can agree that fleeing artificial desperation is a kind of practicality, but I wouldn't call it natural.

I wonder what minds, perspectives, and creative wonders we are losing with our work-or-else model. Our civilization arbitrarily holds a knife to the throat of everyone and calls it natural, while simultaneously priding itself on creating obscene levels of personal wealth for a tiny fraction of the population. This isn't an efficient way to foster and unlock human potential, and if you're tempted to think it is, consider that I never could have written this book you're reading if not for a confluence of rare privileges that sheltered me from our myriad manmade desperations long enough to heal and reorient.

How many would find their joy, their calling, their passion if we, as a society, honored their worth by guaranteeing their basic needs? Do we honestly believe tolerating laziness is too great an evil to risk providing fundamental necessities to all citizens?

Universal basic income could be seen as an investment.

Subsidized childcare, too, as a worthy use of our collective resources to foster a better future for all of us.

What could we achieve if we allowed everyone to reorient from escaping desperation to pursuing happiness?

Our technology and efficiency advances. One day, I hope, our philosophy will advance with it.

You may think, "It worked for me, so maybe people should just step-up and do it like I did."

Good for you.

If you escaped a burning building, would you help fight the fire or argue that the unchecked blaze builds character? Would you worry the lack of choking smoke could encourage excessive laziness?

There is no version of success, wealth, or stoic practicality that makes any human life less fragile, fleeting, or impermanent. It's a truth many of us like to forget, but it smiles at me with stone teeth from beyond my back fence. Even if I didn't live within sight of a cemetery, my depression will not let me lose sight of this basic truth or its wise follow-up question: without kindness or peace or joy or delight, what's the point?

What was it all for?

When I think of leaving this life, it is not the practical matters of living that I regret leaving. I don't think of nutrients or caloric intake, I think of flavors. I think of the people with whom I share my bread. I think of conversations at the table. I think of the funny accidents of family and friendship that become tradition, that accrete meaning like a riverbank gathers footprints. These things are not frivolous. They are the proper foundations for effort and endeavor.

Thinking of my priorities in relation to the end of my life is a useful landmark for navigation. It informs my personal and social philosophies.

One day, my story will end.

All the choices I have made will be frozen like an insect in the amber of history.

I'm building something that will one day be whole and complete. There will be an endpoint. There will be a time when I have built all I can with my choices and actions.

I try to stay present in the moment, but from time to time, it's worth viewing my days through that lens of finality and asking:

What am I building?

Is the project of my life being guided by what I truly treasure?

Am I respecting in life the things I imagine missing most?

Are my choices gathering to me the moments I want and need?

If I am going to be concerned with productivity and practicality, let them be in service to pursuing goals that actually matter. Let me keep those goals front and center.

When it's time for me to go, I doubt I'll regret not being more efficient or rich or famous or respectable.

My imagined ghost will regret not having more time to catch lightning bugs.

I will mourn the end of my little domestic rituals of pine tea and silly secret handshakes with Leslie.

I will miss whimsy.

❖

I once knew a man whose retirement pushed him into a deep depression.

The routine he knew for decades was gone.

The lucrative work that had stood for the bulk of his identity was behind him.

So, he found his mind suddenly dominated by despair and thoughts of suicide.

He felt adrift and without options.

I have never retired, but I have left a career behind. I could relate to some of what he was feeling.

And yet, from the outside, his situation also seemed absurd.

He had a surplus of the two things so many of us desire most.

Time and money.

No, he wasn't twenty anymore.

But he could visit a different country each week if he wanted.

He had the space and resources to learn anything that captured his interest, to discover new passions or explore myriad paths to creative expression.

It didn't matter.

He didn't want any of those things.

He wanted a way back to a life he recognized, one where he knew how to calculate his place and worth with a familiar equation.

I did not want to return to my old career, but I could sympathize with his perspective. I know the longing to commune with past, happier selves.

Being beyond the edge of the map can be overwhelming and defeating.

When you don't have a destination, all paths can seem equally bad.

I had space. I had time.

I didn't have money, but that summer I had a broadening field of options as my depression became more manageable.

But it's become clear to me that the theoretical potential for happiness, especially as measured by an outsider, is not happiness. Not even close.

The potential for a full life, measured in time or space or money, is not a full life.

In a way, the problem parallels the seemingly contradictory fact that antidepressants can increase the risk of suicide. Having more energy is not the same as the skill of shepherding that energy toward happiness, of transforming energy into the beneficial kinds of meaning that serve our wellbeing. Having the space and free time to pursue contentment is miles away from understanding how to pursue contentment.

If I suggested to my depressed, retired acquaintance that he find comfort and meaning in catching lightning bugs, for example, I feel confident he would look at me like I was suggesting he try sailing to the sky on a tide of moonlight.

I would be handing him a kind of useless, hollow positivity.

A slogan on an office poster.

Problems are just opportunities in disguise.

Not because it is impossible to find restorative meaning in a field of lightning bugs, but because lightning bugs aren't the important part.

The important part is knowing how to value visiting with lightning bugs, knowing why you're visiting them and how to find meaning in such things. It's a skill. It's learning to believe there is actual worth in the experience, not as a trick or gimmick, but because of a synthesis between our intentions and the world that can indeed generate very real, very vital, personal truths. The kind of meaning that can heal or sustain.

I adore whimsy, but like magic or nature, the power lies in meeting it halfway.

The energy and will to make that journey, to move toward purpose, contentment, and joy, even toward simple interest, I categorize as enthusiasm. Enthusiasm is our hunger to enjoy life, to employ magic and whimsy, to value pleasures and meaning, to go beyond bare fact and reductive models of practicality.

Enthusiasm is an undervalued resource. I tend to equate it to firewood.

When you were born, your enthusiasm was a red flame atop a mountain of fuel. As you age, as you deal with challenges and disappointments, the fuel burns low. The fire dims.

I don't recall having to purposely cultivate enthusiasm when I was a kid. I was always interested in something. I always had a will to pursue a want. The firewood was just there. I never noticed it was running out because I'd never faced a time without it.

When camping, conventional wisdom says you should gather all the wood you think you'll need for the night, then double it. Gather it during daylight when fuel seems irrelevant. Dead limbs hanging low, sun-dried, hungry for fire. Gathering wood in the dark, in the cold, is not impossible, but it's not an easy or pleasant chore. The night has a way of being longer than we anticipate. The wind can be colder than we predict. The dark beneath the trees is absolute.

Gather the fuel. Double it.

In my unexpected summer of mental health, I had to relearn how to gather and split wood, how to strike sparks onto tinder and coax a flame with gentle breath.

Metaphor aside, this means making room for things that are not your job, things that are not measured in money or necessity. It means relearning motivations not rooted in socially acceptable versions of success-driven practicality.

I have been asked for advice on rediscovering whimsy and enthusiasm.

It isn't an easy question.

But a hint that sometimes helps me is the thought that I knew things about my own interests as a child that I was later incentivized to forget. Those whispered truths from younger me were a key reason Leslie could talk me into walking in the woods when all I wanted was to lie in bed. Younger me did not cross-examine his enthusiasm or his curiosity, so borrowing his perspective was a good place to start when I found my life bereft of interest.

If you struggle to locate your own enthusiasm, there may be a younger version of you lurking in the back of your mind with some ideas worth revisiting.

Enthusiasm will not just feed your own joy, it will connect you with others.

When Leslie and I moved back to Ohio after a stay near Seattle following grad school, we had very few friends in the area. It was daunting. Making friends as an adult feels so difficult.

We began attending science fiction and fantasy writing conventions, an act of pure whimsy and enthusiasm. Writing was our mutual hobby and passion. Through attending panels and seminars, we struck up conversations and met new friends who welcomed us into a larger community built around nerdy, overlapping interests.

I do not think it matters what you choose to care about, just that you embrace it and approach it with willful enthusiasm.

Choose an interest and try pouring a little energy into the choice.

The truth is, once they have left us, ritual, whimsy, and enthusiasm are not impatiently waiting for space to reenter our lives. They need to be invited. They need to be pursued and caught, like a firefly, lightly in the open palm.

You might look silly to passing cars while you fuss over a wayward snake or don a Star Trek uniform in public.

This may be the price you pay to relearn the things you knew had worth when you were a child.

It's alright. Making people laugh is a good thing.

It's worth it.

You're worth it.

Perhaps you, like me, did not set aside your whimsy or enthusiasm because you no longer needed them. I did it, unintentionally, because I was sold the idea that they were no longer of value, that they were childish, inconsequential things. I did it incrementally, little by little, and depression readily occupied every vacated inch as my past joys receded into memory.

There are always many pressures trying to shape what we are allowed to enjoy.

Financial pressures and physical limitations, certainly, but also cultural pressures related to everything from gender to age to the expectations of peers and family. It's easy to get bone-tired and say, "Yeah, yeah, I probably only like the things I'm supposed to like."

"I probably do care most about practicality and productivity."

"I probably am my job."

"Yeah, I guess it is the directionless privilege of youth to care about visits to the woods and catching fireflies."

"Who has time for such frivolous stuff?"

Except such things aren't frivolous, and the forces that need you to believe otherwise want things from you that have nothing to do with your health or happiness.

Colossal life shifts like retirement or divorce or a health crisis have a way of drawing our attention to what is missing most from our lives.

Yes, nature, community, and whimsy stand out as prime examples, but enthusiasm feels like the key to accessing all the others.

The only way to break out of a claustrophobic worldview is to cultivate interest in new things and other perspectives, to imbue the world beyond ourselves with meaning and value. It's the only way to change the furniture in your own mind, to keep the world fresh and interesting.

It's a skill. It takes practice. It happens at the place where choice meets a willful shaping of perspective. The effort absolutely does not need to be perfect to be worthwhile.

Irony. Detachment. Performative and toxic masculinity. Confusing money for value. Confusing practicality for satisfaction. Divorcing success from happiness. Relinquishing the ability to define your own purpose, worth, and meaning. I have fallen prey to each of these snares, and a huge part of my journey toward mental health has involved noticing them and slowly, meticulously, untangling myself (again and again).

I am learning to practice whimsy on purpose.

———

I whisper greetings to trees and I write poems about creatures that don't exist. I nod at the headstones near my fence like a passing neighbor.

I am learning that ritual, as a way to shape and conjure meaning, is worthwhile.

I thank the woods for my walks and touch the last leaf I pass in farewell when my car comes into view.

I am learning to intentionally cultivate enthusiasm.

Tiny pebbles and pinecones sit in places of honor on my shelves where they remind me that my little room is not the whole of the world, that so much more waits beyond my sight.

I am learning that watching a lightning bug crawl across my hand can mean more than any accolade or job title, can easily be the entire point of a day or a lifetime.

Chapter 14

JEWELWEED

ONE OF MY FAVORITE SUMMER WILDFLOWERS is jewelweed, a shade-loving annual that grows a few feet tall. Near my childhood home, they grew in clusters along our driveway. Spotty, orange, conical flowers perched high on thin, leafy green stems.

When my mom was teaching me wildflower lore, jewelweed captured my imagination for two reasons. First, jewelweed is also known as touch-me-not because its ripe seedpods will burst if prodded, shooting seeds a surprising distance. Second, its sap has wonderful anti-itch properties that are good for sunburn and poison ivy rash.

These were, and are, wonderful things to know.

A lovely plant with both a party trick and a healing touch.

I often reached for its sap to soothe stings and rashes.

Is jewelweed sap the most effective medicine for itch?

No, certainly not.

Yet to a scraped-up kid with mud-caked boots, it was often my preferred medicine because it was there. No getting cleaned up to go inside. No asking for help digging through the medicine cabinet.

I wasn't about to risk being told it was time to stay in, not for an itch.

It was there. It was magic. It felt like secret family knowledge and like the woods itself was interested in taking care of me.

What could be better?

I do not doubt that some of the benefit of jewelweed came from placebo effect, but that didn't matter.

If the itch feels better, then the itch feels better.

Sometimes, the best medicine is the one we are willing and able to reach.

It doesn't need to be ideal in order to be good.

❖

There is a lot of doom in the air these days.

Doom well beyond a poison ivy rash.

Doom arising from legitimate loss, tragedy, and risk.

Danger and injustice.

Ecological. Personal. Societal.

These threats and tragedies can complicate our discussions of mental health, especially in light of that slippery old truth that depression defends itself. Many other forms of mental illness do as well. There are reasons to be sad and anxious. That fact is neither a mental illness nor, in CBT terms, a thought distortion.

So, a central question tends to arise.

Is it reasonable to feel peace or happiness in such an imperfect world?

Is it moral to look for contentment when so much worthy work remains undone?

Are our very lives a liability to the planet?

Not long after I began to find an audience for my poetry and podcast, I rented a PO box so I could receive mail from listeners and readers. I love physical mail. I love the sense of human touch in the paper. The intimacy of ink.

One of the first letters I received was from a woman I'll call Mary.

Mary complimented my appreciations of nature, then spent the bulk of her letter condemning humanity as a disease on the planet.

She moved from a broad condemnation to regret aimed specifically at her own existence as a creature that harms the world simply by consuming food and participating in a cultural context rooted in exploitation and destruction.

She said that the Earth was not going to gain a positive return on investment for the food she ate and the air she breathed.

She found that the hopeless despair of her situation made every appreciation of nature bittersweet.

Understandable.

In her worldview, she was the villain. The condemner and the condemned. She was complimenting the house while lighting the curtains on fire. For Mary, the only moral option available was self-hatred.

She was an accountant of ethics and found that her own books were hopelessly in the red.

Familiar with the culture that encourages us to think of our entire lives as business ventures, I found her phrasing, "return on investment," stuck with me.

I wrote back, trying to be gentle in my challenge, fully aware that any disagreement might register in her perspective as moral corruption in me. In offering another viewpoint, I would be an apologist for humanity. Worse, I might seem like I was arguing in favor of the status quo.

I did my best.

She had plenty of points I could affirm.

Yes, we could do so much better in our efforts to preserve our natural world.

Yes, I think vegetarianism is a meaningful, selfless way to contribute and is worthy of pride.

Yes, governments and societies have done immeasurable harm in terms of pollution and loss of biodiversity.

But I needed to express where our worldviews parted ways.

We cannot take full personal responsibility for global outcomes beyond our control.

Humanity is not one cohesive group to be dismissed or condemned.

Every individual is worthy and fundamentally deserves joy and contentment.

The world is not a business, and no living creature is a product or an investment.

Mary thought of herself and her species as a sort of existential mistake, a planetary miscalculation.

But existence just doesn't work that way.

Life doesn't work that way.

Earth is not trying to achieve anything in particular and, as such, is not reaching for peak efficiency or any single win.

You aren't an investment and can't be evaluated as such.

Mary was taking on a level of responsibility (and blame) that would suggest that she was in control of not only her civilization and cultural context, but the nature of her existence.

She was confusing her insistence that she should be in control with actual control.

She was falling prey to a thought distortion, taking on an unrealistic sense of culpability.

But she isn't responsible for events and patterns beyond her control.

And she doesn't deserve condemnation for circumstances beyond the scope of her choices.

For some of us, it's easier to accept blame for things beyond our control than to acknowledge the stark limits of our agency. I certainly felt this way for many, many years. It can feel better to accept self-condemnation for willful negligence than to feel powerless or irrelevant through coming to terms with our actual sphere of influence.

I suspected that Mary might relate.

She not only took responsibility for all of humanity's mistakes, but seemingly for nature's mistake in creating beings that both consumed other lifeforms to live and had such massive impacts on climate and ecology.

Mary was not consulted when our current agricultural processes were developed.

Nobody asked for her guidance in shaping modern capitalism or the industrial revolution.

She didn't invent the processes of life that brought her here nor can she edit them.

She certainly didn't choose what sort of creature to be.

The blunt truth of our living world is that unless you photosynthesize, you're living off of other organisms.

You didn't create this arrangement. No one asked you to sign off on this fact.

Life exists in myriad forms and its existence is its own justification.

Nature doesn't ask for justifications.

People do.

It's okay to think otherwise, to believe that life is here for a specific purpose. Just remember, that sort of thinking is about human interpretation. Making a *why* from a *what*.

Making meaning.

Interpretation.

Our meaning matters, but it becomes murky when we believe we've decoded a fundamental flaw or essential immorality woven into the fabric of nature.

Is it so easy to second guess the unknowably ancient systems through which our own bodies, minds, and thoughts arose?

So, when Mary tells me she sees herself and her species as a plague on the world, I don't sense I am learning something about nature or humanity.

I sense that I am learning something about Mary.

Primarily, that she and I disagree on a key principle.

An intrinsic right to self-love.

I believe that every human being living is as natural as a spreading oak shading a river valley.

As natural, as worthy, and as deserving of life, dignity, and consideration.

Every person is as much a part of nature as that oak.

Every person has as much right to stand tall in the sunlight as that gorgeous tree. Equal in value. Equally "correct" for our planet. Autotroph and heterotroph.

Beyond that, a person is not a cumulative tally of their species' perceived harm.

I don't even believe a person may be reduced to a tally of their own harm.

We are not our carbon footprint nor our water consumption.

That oak and each and every human child arrived here through the same universal processes and are heirs to the same nourishing mysteries of life, growth, change, and impermanence.

It is undeniable that humans sometimes make needless, cruel choices and perpetuate systems that harm ourselves and our ecosystems.

No question.

In other words, people are fundamentally imperfect.

But that doesn't mean we're suddenly outside of nature.

It doesn't mean we are no longer worthy in the same way as our kin, the trees or the squirrels.

There is no "best" way to be alive or "ideal human model" for the planet. We are temporary. The planet is temporary. What is ideal for one species is sometimes less than ideal for another. The very notion of "best" is a matter of interpretation, crafted meaning, and I am suspicious of any meaning that does not make room for self-love and kindness, that does not acknowledge that humans are as natural as any other creature of this world.

Can we do better?

I certainly think so.

We have many, many models of mutually beneficial relationships in nature.

We can look to business arrangements like fruit in exchange for seed distribution or nectar in exchange for pollination.

We can learn a lot from such teachers and would do well to emulate them, the way their livelihoods enrich rather than deplete their environments.

When I point out what is beyond our control, this isn't me suggesting we all take our hands off the levers of our lives and shrug.

Our choices absolutely have weight. There is balance to be achieved and I think it should be our task to seek it.

Whether we're discussing institutional racism or gun violence or the carbon emissions driving climate change, these are human problems that intrinsically must have human solutions.

The solutions must be our responsibilities.

———

This, of course, means investing ourselves in awareness, understanding, education, collaboration, and action.

And yet, there is no possible way within the limits of a human mind, body, and lifetime to treat all of the very worthy issues of our planet and species equally, or to act on them all with perfect clarity of knowledge and surety of purpose.

There is no certainty of moral perfection or purity to achieve in this huge, complex world.

There is no unassailably "correct" set of priorities and actions.

We all have our own set of privileges and priorities.

We all experience certain challenges, dangers, and injustices in differing measures.

We all, in short, lead different lives.

I, naturally, tend to see things through the lens of my own mental health struggles.

This is one reason my worldview must stubbornly start with a premise that, no matter who we are or what we do, we are each a worthy part of nature and deserve to feel contentment.

Mary's perspective led her to the idea that her entire existence is a glitch, a mistake, a blood vessel fueling a cancer cell.

All-or-nothing thinking.

We blew it.

From a baby you've never met noticing her fingers for the first time to the invention of the roller skate, the universe really botched it with us.

Human-driven extinction impoverishes the world, so our grandmother's chocolate chip cookies turn to ash in our mouths.

A warlord and a nightshift nurse, all part of Team Mistake.

All part of the same cosmic crime that is humanity.

How very cut and dry.

But weighing the value of any lifeform's fundamental existence or any person's limited life-minutes isn't like finding the most efficient way to organize your pantry.

What would be your criteria to judge?

Life isn't a spelling bee.

We are not here to "win."

We are here to experience and to touch the world with our own choices and values as best as we are able.

It is dangerously easy to fall into all-or-nothing models of thinking when considering morality and ethical action.

My depression constantly asks me if I really deserve what I have.

Jarod, why do you get a roof over your head?

Jarod, how do you justify sitting down and resting?

Jarod, another cup of coffee? Isn't that an egregiously frivolous expense?

Many viewpoints less extreme than Mary's also carry the weight of endless guilt for failing to be enough of a force for good in the world.

The trap of a purity obsession.

Are you doing everything possible to serve your own values?

Let me answer that for you.

No.

It's not because you are lazy or immoral.

It's because you are a living creature and no part of your being is perfect or even consistent.

This is not because you are broken, but because perfection is a fiction. It's an abstract, subjective concept that dances away like a heat-mirage when you reach for it. There is no perfect version of ourselves or our world to achieve. Because perfection does not exist, it cannot be the goal.

You will never be good enough to justify the air you breathe, the food you eat, and the space you occupy.

Not if you believe such things are justified through meritorious action. Not if you don't accept the idea that the simple fact nature and happenstance brought you to this moment means you are supposed to be here, means that you deserve to be here in a way that honors your inalienable worth.

Our empathy and our sense of morality compel us to act.

Frankly, I love this about us.

But what is enough?

How do we extend empathy to ourselves, holding dear our fundamental need for joy and contentment, while also honoring our pull to serve others and the planet that nourishes us?

As I have said, a cornerstone of mental health is acknowledging what we do and do not control. We have to start there. We have to be honest with ourselves.

We are not weak.

We are also not omnipotent.

We have influence over many aspects of our lives, but we primarily control ourselves.

This is one reason why it is damaging and disingenuous to take responsibility for global outcomes. We do not control them. They are beyond our power.

We do not, as individuals, steer the ships of social or ecological issues involving billions of other human lives, billions of other perspectives and personal modes of meaning. Complex contexts and cultural inertia.

Acknowledging this can be painful.

Painful because we're immersed in narratives of both global doom and of individual power and purity of purpose. The lone superhero. The driven, solitary genius. Figures that are attractive because they are small enough for us to conceptualize, but big enough to do it all on their own.

Yet, in real life, it's never one person who does the work of global change.

So, honesty compels us to recognize that we cannot take responsibility for global outcomes because we do not own or control them.

This doesn't mean we have no responsibility at all.

We have the responsibility to do what we can, while honoring our own natural worth. The worth that commands us to save room and energy for our own happiness.

Picture a stone, a towering boulder with a lake-sized shadow.

This boulder is the global issue or outcome that is troubling you.

You cannot hope to move it on your own.

Find a way to make peace with that.

You did not build the boulder and you cannot move an object of that scale by yourself.

You could certainly choose to break your body against the stone, many have, but your shattered bones will not shift all those tons of rock.

Daunting, yes?

And yet, boulders of this size are moved by human hands all the time—many hands working in concert.

This is our obligation. To lay our hands lightly on the things we can influence and press in the direction of our values. If enough people press in the same direction, the ground will rumble and the boulder will shift. We can do this, our part, while acknowledging our vital constraints, namely the necessity of honoring our own limitations and our unshakable right to mental and physical wellbeing.

There are many boulders that deserve your touch.

Many will not all receive it.

And you cannot control who will join you in pressing on your chosen boulders.

We tend to focus on those issues close to our lives, the struggles we know best, the boulders within reach.

This is not a reason for self-condemnation.

Humans are small, limited, temporary things.

We cannot push against all the worthy boulders equally and simultaneously.

This truth is one more thing that is simply outside our control.

Be disappointed by this.

Don't be damned by it.

The world is a worthy place, full of worthy choices, and it can be difficult to know which to choose. But face this difficulty in the knowledge that we deserve to feel moments of happiness and peace. We are each a part of nature, part of this worthy world, and we deserve care and consideration. Even in a world where our virtuous work will never be complete, we deserve to put effort toward our own mental and physical wellbeing.

It is true that some people don't have the privilege of either mental or physical wellbeing.

This is one of the boulders we press against. It's not a reason to forego happiness yourself. Breaking your body against the stone doesn't shift the stone.

Understand that the good and evil in this world are happening concurrently. This has always been the case. Making time for the good is, itself, one way of defying the evil.

Those are big words.

Good and evil.

Big boulders looming over our landscapes.

They are worthy of attention.

But they aren't the only things worthy of attention.

The quiet, gentle, tender moments deserve as much or more attention as our dramatic, strenuous, industrious moments. Lives are mostly quiet, small, fleeting things, and it feels like a mistake to spend them wishing instead to be huge, enduring, planet-shaping things.

Mary is not the architect of the universe.

Neither are you.

Neither am I.

Context.

Honesty.

Acceptance.

We are worthy parts of an amazing world. We have meaningful choices to make. Our limitations do not negate our worth or right to happiness.

Your job isn't to move the boulder, just to push when you are able.

Cultivate love for what is and hope for what could be, all while sparing a bit of awe and gratitude for the massive, unknowable, unlikely forces that culminated with your own singular presence on this planet.

I sent a letter in reply. I don't know if sharing my perspective helped ease Mary's burden. I worry that she needed something stronger, something from the medicine cabinet, when what I had to offer was jewelweed.

She did not write back.

I posted some of my response online, hoping it might soothe others adrift in self-loathing and falling short of perfection and purity.

I did what I could.

I pressed against the stone.

That feels worthwhile to me.

❖

Regional and common names for plants and animals often tell you something about the people who did the naming.

In terms of jewelweed, the name touch-me-not suggests I'm not alone in thinking those exploding seedpods are one of the coolest things about the plant. Yet, insomuch as a name is meant to capture the essence of a thing, how much is an individual jewelweed plant defined by a dramatic little snap of seeds tumbling into the air?

Our eyes are drawn to the dramatic, the showy.

We value action.

Humans find conflict and destruction compelling.

Our brains are attuned to threat detection and assessment.

This is a tendency worth noticing, especially as we discuss our own drive toward moral action in a flawed world, our own need to press on those towering boulders.

As I continue my lifelong work to understand my own mind, I've been thinking about the difference between constructive and destructive mindsets. The difference between fighting and building. Opposing and cultivating. Condemning and celebrating.

It's no secret that we need our destructive mindsets to oppose injustice, fight those who do harm, and speak out against abusers, bigots, fascists, and others who seek to injure, dominate, and exploit. Yet I think it's clear that what makes the world (and the inside of our skulls) a joyful, livable, sustainable place to dwell requires constructive mindsets. Building shelter. Building communities. Food. Art. Education. Childcare. Science. Medicine. Fun and leisure.

Enthusiasm.

Whimsy.

Anger may hold our attention, but it rarely feeds or cares for anyone.

We know that many online spaces capitalize on outrage, fear, and anger to hold our attention. It certainly works on me. This tendency pulls me toward my destructive mindset.

I will catch myself doomscrolling and realize that my jaw is clenched and my heartrate is elevated.

I know I'm not alone in this.

So I advocate intentionally steering toward our constructive mindsets. However, like seeking happiness in a flawed world, it's not so simple. Both reason and a respect for my own sense of wellbeing tell me to intentionally cultivate a constructive mindset, but in doing so, I must make an active choice to turn away from unaddressed threats, problems, and injustices that may well require a destructive mindset. How do I honor my need for peace while standing in opposition to danger and injustice?

This dichotomous, all-or-nothing thinking is the same mental pitfall that says we must lend our efforts to all moral issues fully and simultaneously or we cannot be moral beings.

The idea that we must dedicate ourselves solely to one or the other, constructive or destructive mindsets, is a false choice.

We need to make room for both.

And finding that balance requires me to acknowledge that there are vast, sophisticated tools, algorithms, and financial interests pressing down on the "destructive mindset" side of the scale, keeping me locked in anxious engagement.

I often have to remember to activate my constructive mindset.

My job involves a lot of time on social media, so I am often exposed to the destructive influence of online spaces, where anger and outrage make us more lucrative products for those who monetize attention online.

If the content I'm consuming is positive or neutral, the science says I'm more likely to wander away. If the content threatens to launch me into fight or flight mode, I'm more likely to stay focused on assessing the potential danger.

The problem is exacerbated by the abstract placelessness we feel as citizens of the internet and as people who have been cut off from our physical contexts by recent factors including the coronavirus pandemic and more general isolating aspects of modernity. We, in a sense, become inhabitants of social media. It overwrites our physical spaces and becomes our environment, an environment with an active interest in keeping us angry and fearful. Threat as a place. The natural environment, the one we evolved to inhabit, is not an intentional construct for an intentional purpose. Social media is. That distinction is one reason we are so ill-adapted to face this new threat to our sense of self and our sense of peace.

No wonder so many of us walk around ready for battle.

When we seek ways to exercise constructive mindsets, to find meaningful work centered on hope and pleasure, the deck is stacked against us. Yet mental illness already stacks the deck against my hope and pleasure. Social media is not half the monster my depression is. And monsters can be teachers.

I am well acquainted with insidious internal pressures trying to steer me toward hopelessness. I am well acquainted with having to make a conscious effort of will to turn toward positivity, to notice pleasure, to go outside, to recognize and acknowledge when my anger or despair stems from forces beyond my control. Social media may have made rage and fear into a business model, and they may be very subtle and sophisticated about it, but they aren't headquartered inside my skull.

I read what others are posting.

A lot of folks seem fully invested in the idea that violence is the answer to many of our problems.

They are done pressing on boulders. They are ready to bring out the explosives.

Violence is sometimes necessary. I've seen armed white supremacists demonstrating on a streetcorner less than two miles from my home and, despite the documented statistical risks involved in keeping guns in my house, there are reasons I'm not content with allowing right wing extremists to hold a monopoly on possessing firearms. Still, I find the hand of social media, in concert with perhaps a bit of toxic masculinity and our human interest in the dramatic, artificially inflating the role of violence in the imagination of many who crave change.

Revolutions may need to fight, but they also need to feed people, to make life worth living, to present a vision of a world that feels worth risking fundamental change. Destructive mindsets have their place, but we miss the point when we let them define our goals and identities completely.

I intentionally seek out things that make me feel hopeful.

I stubbornly allow for the idea that many of my fellow human beings are good, are smart, are worthy, are interesting, are enriching the world with their presence.

I recognize that I am susceptible to manipulation and that social media isn't a trustworthy representation of reality, that its reason for existence has very little to do with communicating news or truth.

I adopt the self-care stance that in this flawed, complicated, temporary world, the local trees and birds are also deserving of a portion of my undivided attention and that giving it to them is neither a surrender to evil nor an immoral act of self-indulgence.

Sometimes, the portion of nature for which we are best positioned to care and preserve is ourselves.

Yes, we should oppose evil.

We should take action.

We should do good works and press against looming boulders.

But if you find yourself living in a state of constant dread or burning rage, I want to recommend that there's a healing balance to be found between destructive and constructive mindsets. The destructive mindset has many champions and many loud calls to arms. Often, however, we have to be intentional advocates of a constructive mindset within our own lives.

Nurturing ourselves and others is rarely a dramatic, showy, sexy endeavor.

It rarely builds wealth or grabs headlines.

Constructive mindsets are, like the phytoplankton producing most of our oxygen, quietly and unobtrusively responsible for our continued survival.

War isn't growing crops or changing diapers.

War isn't teaching or healing.

We are all worthy of seeking balance. Of finding hope. Of rediscovering our place and peace. Of forgiving ourselves for what we do not control. Of allowing ourselves to be simple, natural animals enjoying the beauty of this flawed, lovely world.

If, at the sight of jewelweed's little, orange flowers, you obey the directive of touch-me-not because one week out of the year a tiny pop might startle you with a shower of seeds, you miss the healing sap that flows all summer long.

We do not need to let destruction own all our worthy narratives.

We do not need to let the existence of unhappiness banish all happiness.

We do not need to be defined by all-or-nothing thinking, to solve everything or nothing, to lose ourselves in the strictures of unyielding myths of moral purity.

You are enough.

Your modest efforts are worthy.

Gentleness, care, and compassion have preserved far more life than violence and rage ever could. No living person would have survived infancy without years of strenuous, stubborn love and care, yet summer blockbusters rarely celebrate things like childcare. Destruction makes money and grabs attention, but it doesn't make life worth living.

Take time to set aside anger and embrace your constructive mindset. One of the most vital things for you to build and nurture is your own inner landscape. It is hard to bring peace to others when you do not grant it to yourself.

It is neither immoral nor irresponsible to seek joy and contentment. Reaching for goodness is how we affirm our own intrinsic, natural worth and recognize that life deserves the effort. A flawless world cannot be part of our criteria for allowing happiness into our lives. Jewelweed is not the perfect remedy, but it is within our reach and that makes all the difference.

Chapter 15

VIRGINIA OPOSSUM

SEPTEMBER ARRIVED and looked very much like August.

Hot, muggy, and buggy.

The weather was rich and sweet and, in the classic character of summer, seemed to insist that it would be so forever.

Cold felt like a memory from another lifetime. The acute pain of my depression, the desperate, constant thoughts of escape and suicide, likewise felt distant.

It's scary how quickly the present becomes the past.

Summer spoke in a deep, hazy voice and said, "I am here to stay."

The meadows that skirted my beloved woodlands said something different, speaking in a language of flowers. Thistle and goldenrod and red clover. Cattails shifting from sleek brown to shaggy white. Other flowers too, ones I knew by sight, but not by name. Blooms I knew would still be in the field, side by side with the woods, when the oak leaves were turning brown and the maples red.

Fall was coming.

I bought a cord of wood and began having backyard fires at dusk as the temperature dipped.

The lightning bugs were gone, but bats still fluttered in the twilight.

One evening, I went out well after sunset. It was true dark, but on a whim I decided to make a little fire. I stepped out back and there was a plump, gray Virginia opossum sitting on my back porch railing nosing at the birdfeeder.

I will ask my taxonomist friends for their forgiveness for my using opossum and possum interchangeably. I am a product of my region. Technically speaking, possums and opossums are very different animals. Colloquially, they are not. I was an adult before I ever heard anyone say "opossum," pronouncing that initial O.

She swung her snout in my direction, in the slow, careful way possums always seem to move.

I love possums.

I love that they're my continent's only marsupial.

I love imagining what strange happenstance brought their ancestors to these shores.

I love their sleepy movements.

I love that they seem to be half storm cloud and half child's doodle of a deep-woods cryptid, clever clawed fingers and that naked, prehensile tail.

I love the fairytale way an opossum mother becomes a bus for her children.

They are gentle, omnivorous, beneficial creatures that will quite literally play dead rather than harm you. Still, I once found an opossum skull in the woods and the image of it stays with me. All those sharp teeth inspire respect.

I don't speak possum.

So, I told her I meant no harm in the language of deer.

I sat down.

In surprisingly short order, the possum turned away from me and went back to munching birdseed.

I spoke to her in a soft voice.

"It's nice to see you. Are you finding anything good to eat?"

She ignored me so completely. I decided to go inside and get her a snack.

I came back with unsalted, shelled sunflower seeds. I walked to her in a wide arc so she could keep an eye on me and leaned against the railing about seven feet away. I assumed she would run for it and I would leave the seeds for her to find later.

She did not.

She stayed motionless for a moment, then began sniffing in my direction.

I reached toward her and dropped a little hill of seeds between us.

Soon, she approached my offering.

There was a flash from our sliding-glass backdoor.

Leslie had noticed what I was doing and had taken a picture.

I smiled at her, then swallowed.

What I was doing was irresponsible.

I knew better.

You shouldn't feed wildlife.

This wasn't like the ratsnake. There were no extenuating circumstances that justified my behavior. I was unwisely acclimating a wild animal to humans, and I was doing it for my own entertainment. I had no justification.

I wasn't thrilled about her documenting the moment.

My dog Atticus woofed and clawed the glass.

The possum fled, climbing off the railing with surprising speed.

I turned and shouted at Atticus, while Leslie pulled him away.

There, steps from the porch, laid the possum on her side, mouth slightly ajar.

She was playing dead.

Playing possum.

Defensive thanatosis.

It was a fascinating and unexpected sight. Unexpected because typically possums only perform this behavior when they can't escape, but apparently all that sudden commotion was too much for my gentle visitor.

I sighed.

I had made a big mistake.

Atticus was a rescue from a shelter in Appalachian Ohio. The folks there told me he came to them "from the middle of nowhere." In response, I thought but did not say, "We already *are* in the middle of nowhere."

Atticus is a sweet dog that presses his head against me when I'm sad and regularly climbs into my lap for a hug even though he weighs 80 lbs.

On the worst of my depression days, he stays by my side without being asked.

Having Atticus around means I rarely cry alone.

He was also almost certainly feral before coming to the shelter.

When we brought him home, he had never seen stairs.

He didn't seem to recognize food or water bowls.

He happily asks to sit outside in deep snow or in the middle of a thunderstorm.

Sometimes, he seems surprised he is allowed to come back inside.

He is dotted with all sorts of old scars.

And, if I don't shoo the squirrels far enough from the house before letting him out into our fenced backyard, he brings me one of their bodies. At least one.

I once had to scold him before he could catch a starling in flight, having watched from the back porch as he gained on the bird flying low across the yard.

He is a kind, loving dog whose prey drive probably saved his life once upon a time.

Playing dead would not protect a possum from Atticus, and I had just given one a reason to revisit my backyard.

I did what I could to correct my mistake. I brought in my birdfeeders and resolved to be extra vigilant before letting the dogs out. I kept the floodlights shining on the backyard at night as a deterrent. I worried it wouldn't be enough.

And it wasn't.

Nature is not a toy.

The lives of other creatures are not playthings or ornaments for our egos.

A few days later, on a cool, September morning, I walked out back in my slippers to see how the mists lay among the headstones. I found the possum's chewed body by the back fence beneath a blue spruce.

Ten to one Atticus was the killer. But I couldn't avoid thinking that the responsibility was mine. No, I do not control all of the events in my backyard. I did not control the possum. I did not control Atticus. But my choices influenced the situation, and it seemed my influence had proven deadly. I knew better than to feed the possum and I did it anyway.

I set my slippers aside on a cinderblock, retrieved a shovel from my shed, and, barefoot, I buried the beautiful creature in the soft earth near the fence, extending the territory of the cemetery another few feet. I stacked stones on the grave to thwart digging paws and said a few words of apology and farewell.

I couldn't help wondering if the possum was playing dead when it died.

I tried and failed to resist imagining it.

It hurt.

I let it.

❖

I have a tendency to play dead.

When I feel threatened by dangers, real or imagined, I preemptively cut myself off from life.

"You can't fire me, because I quit."

When I was younger, I was afraid of cities, so I did my best to stay away from them.

My mom was afraid of highways and tornados and I joined her in those fears.

I was afraid of crowds, so I found excuses not to go to concerts or events with friends.

Fear of loss or embarrassment prompted me to keep my head down in sports or in the classroom.

I could happily sit in pitch black woods listening with interest to the scuffling sounds of unseen creatures in the dark, but I was afraid to talk on the phone for fear of sounding silly.

When I first went to college in Athens, Ohio, a town somewhat infamous for its parties, I spent a great deal of time in my dorm room. To me, bars meant violence. Crowds meant violence. Parties meant violence or potential embarrassment. I often felt like my friends were naive for failing to recognize the dangers and that, if I went out with them, I would be called upon to stay vigilant and to protect them.

So, I stayed in my little room, rushing to bury myself alive before anyone could do it for me.

I could, I thought, at least maintain control.

I could be safe. A dangerous concept.

Depression, and acquiring the skills to fight it, has taught me many things. How to differentiate between senses and feelings, choices and beliefs. How to build homemade enthusiasm. How and why to choose whimsy. How to ground myself in the present and in nature. How to reject myths of purity and perfection. How to respect the craft of making meaning. How to find happiness in a flawed, impermanent world. How to be enough.

I have not learned how to banish fear, and I'm not sure I want to.

I am still, frequently, afraid. The difference between my fear now and the fear I held when I was younger is that I no longer feel obliged to act on that fear. Fear is a feeling, arising from a thought, sometimes a perception. Yet it isn't an action. It isn't a mandate. It doesn't get to decide what I do next.

I have not conquered fear, but I have found that when I defy it, again and again, it is diminished. Its demands become optional. The fear still comes. It still barks orders, but I no longer reflexively obey.

Every time I am frighteningly vulnerable in my writing or attend a party with strangers and my world does not end, the fear loses a little more ground to experience. I take another step away from defensive thanatosis, dying so that the world will not kill me.

Playing possum feels like it works, until it doesn't.

Maybe staying away from cities and crowds and parties spared me from some danger or discomfort. I'll never know. I can, however, take an educated guess at what it cost me in terms of unmet friends, missed experiences, and absent memories. I can also say, definitively, that my fear did not keep me safe.

I am not safe.

I never have been.

None of us are.

We are vulnerable, temporary creatures and there is no careful choice that will change that.

We focus on safety because we fear pain or loss or discomfort, reasonable things to avoid, but the foundation of my general fear is an aversion to change. All change.

Often, a false premise lies at the heart of what we seek to protect. We want to preserve an unchanging version of ourselves, an impossible version, and we want to do it forever.

I have humored endless fantasies of power and immortality, fantasies of safety drawn out to their extreme conclusions, the power to be unaltered by anything beyond my choice, including the passage of time.

Living on my own terms.

Safe forever.

These are perfectly natural impulses and I have come to distrust them entirely.

Safety is not my friend.

This isn't me saying we should all go play in traffic, but I am saying we should not avoid being near traffic for fear of what might happen.

I am a person who has stood at the edge of death more than once. Playing dead did not stop the dog's bite. It did not save me from sleepless, suicidal nights trying to empty my mind and be still because simple existence was pain.

Playing dead did not keep the gray from my beard or the suffering from my perception.

All it did was allow me to pretend that I was fully separate from a world that I couldn't predict and did not control.

And yet ...

Losing what and who we love feels like a reasonable thing to fear. Some decisions may preserve us while some may lead us to pain or destruction. There are many things in each of our lives worth protecting.

So, how can we smile and accept a reality that will, eventually, take away everything that we hold dear?

In such a world, can we avoid being defined by sensible fears?

How can we possibly look on our own unquestionable vulnerability with anything but distrust and terror?

In all honesty, some days I can't.

More interestingly, some days I can.

Consider this. The impermanence we fear, the impermanence that seems endlessly poised to take from us, also gave us everything that we know and love. In a static, unchanging world locked into a rigid forever, none of us, nor the things we hold dear, ever would have arisen. In such a reality, there wouldn't have been the space for us to arrive.

It helps to think of the universe as an event rather than a place.

That shifts the goal from owning and controlling to experiencing, from possession to participation.

This existence is not a patch of ground to build a fence around. It's a fireworks show. It's a birdsong among the trees. It's a sunset turning the ocean into a sheet of gold.

It is evident to me that we do not possess life in any real sense. You cannot possess something you were given without asking, something that will one day depart with or without your permission.

We don't own this life. We witness it. We participate in it. We spend our parcel of choices as best we can. We are both audience and choir in the music of the universe.

Somehow, this feels both true and absurd. Yet, on the days when my meaning-making skills are sharp and I am well-awake within my own mind and power, I can see the kinship between the absurdity I feel in the bare facts of existence and the whimsy I actively choose to nourish my life.

My depression often feels like weaponized boredom, like a smothering blandness in all things. The full and final extinguishment of interest and enthusiasm. Both the absurdity and the whimsy of this world remind me that neither my happiness nor worth is granted by some sober council of human rectitude or the conventional practicality police.

When my life seems like thin, room-temperature gruel, I reach for a handful of science and a handful of magic. I think of the big bang. I think of how the universe continues to expand. I recontextualize. We live in an ongoing explosion. Somehow, that same explosion led to this thought, led to this writing. When an explosion explodes hard enough, dust wakes up and thinks about itself. I reach beyond my little Ohio skull and my little Ohio walls and my little Ohio town.

Somewhere, there are orcas.

Somewhere, there is a living bristlecone pine that stood when the pyramids were built at Giza.

Somewhere, an unnamed creature swims in dark waters and has never witnessed an electric light.

This is where we live. This is our world.

I am a wild, inexplicable thing.

I am from and on and of a world of wild, inexplicable things.

I am a part of cycles and systems beyond human understanding, predating my birth and postdating my death.

So are you.

So is that robin you heard singing while walking to your car.

As I write this, a blue whale's heart is circulating nearly ten tons of blood through the largest animal we have ever known to exist.

Not the largest animal *now*.

The largest animal *ever*.

As I write this, the largest organism on Earth, a fungus beneath the soil of Oregon's Blue Mountains, quietly occupies over 2,300 acres.

As I write this, we share the planet with immortal jellyfish and snapping turtles that can live without inhaling oxygen for six months.

As I write this, there exist bacteria that can eat radiation and bacteria that can eat plastic.

This isn't a world for safety.

Think bigger.

This is a world of miracles.

Sometimes, I feel as if my craving for safety creeps in to fill the space my enthusiasm or awe once occupied when I allow them to atrophy from disuse.

The trick is to coax the awe back and prompt my thoughts to take root, once again, in the grand truth that we are a part of this amazing world and a colossal, mysterious universe.

This, as ever, requires choice and intention.

Blue whales and snapping turtles aren't going to knock on our doors to brag.

We must choose to notice.

Thankfully, when I have trouble summoning my awe, or lack the energy to seek it, I can lean on words, that wonderfully portable, sharable magic poised to rescue me.

So, I offer you a pep talk, a version of the kind I often give myself and have used to connect with others:

> If you can make peace with the unlikely fact that squid the
> size of school buses patrol the dark oceans at a depth that
> would crush you to paste, then I have faith you can also

make peace with the unlikely fact that you are worthy of all the happiness you can imagine.

Being afraid is natural; it is our mind's method of protecting us, but don't lose sight of what you are.

You are so much more than a body to be kept safe from harm.

So much more than fear or instinct.

So much more than the sum of your parts.

Not safe, but too rich and strange to be called "fragile."

Zoom out.

The first living cell on Earth was a spark.

It ignited a chain reaction that thundered across billions of years, an evolving blaze of lives and needs and firsts and lasts. The whole of that raging fire of history and happenstance burns inside you right now. It is beyond amazing that you are here reading this.

You should be impossible.

Living is a rebellion. An old conspiracy of elements to organize and awake to sensation.

You climb hills. You swim against the current. You chip away at the expected.

Everything you have ever done is unlikely and is forever a part of the history of the universe.

You are rogue matter.

Your every active choice raises the flag of resistance.

It would be an understatement to call you "rare."

You are so much more than just rare.

The universe is mostly nothing, but there is a tiny percentage of conscious rebels like you.

I don't know where you get the courage to be so unique, but I'm impressed that you choose to be here, caring for a living body and making active choices in defiance of the norm.

You are exceptional.

Today, you did things that humans 50 years ago wouldn't believe and 500 years ago would struggle to imagine. You know the parts of an atom and the topography of the internet.

You ponder death and make art.

Even in the context of human history, a very unusual history, you are a remarkable person, worthy of notice.

Do you realize what an adventure you are on right now?

The dangers you have already conquered to be here?

Do you ever stop to give yourself credit for your achievements?

You are many incredible things, but you are not safe.

Your birth was reckless.

Lightning strikes without reason.

Countless simple bodily mishaps may be fatal.

To live is to collect risk like a chipmunk gathers seeds against the cold.

But risk is not how we measure our worth.

Your primary goal is not safety.

It never was.

Nothing in this wide world is sheltered from change.

Instead, concern yourself with becoming the fullest version of you in this moment.

Not forever. Just for now. This moment.

Every part of you is on an unbelievable journey. The conscious and the unconscious. The simple existence of your mental and physical being is extraordinary on a universal scale.

Let's take inventory of some basic facts.

The atoms in your body have traveled farther than you can conceptualize, yet here you are.

Most planets don't support life, yet here you are.

Life on Earth didn't start on land, and yet, here you are.

Your individual genetic map is profoundly unique, yet here you are.

Don't discount all the impossible things you have already accomplished without a thought.

What impossible thing will you do next?

What impossible thing waits for you to learn, here in this vast and unlikely world?

Let's set aside goals and accomplishments.

Knowledge and improvement are admirable, but hold firm this central truth: there is no "best version" of you.

There is only this version, which stands head and shoulders above all the hypothetical versions you can imagine by virtue of being real and here in the face of everything you have already overcome.

Zoom in.

Feel the shape of this moment, the edges of your perception.

Come home to your body, your senses.

Feel your breath, feel your control.

In and out.

Each breath a conversation with the wilds, with the distant forests and phytoplankton sustaining your life.

The oxygen, the words, coming home to your blood, your heart.

Touch your own skin.

Imagine you are new here.

Exist for a moment as life wondering about itself afresh.

A simple human touch.

A new and ancient machine, a unique individual, connecting with itself.

You.

Engage each of your senses.

What is near you?

What do you have the unique privilege of perceiving right now?

This span of a breath, this sliver of time, this new kind of now, is as real and vital as all its past and future kin.

Grant it the meaning and weight of your careful attention.

Rest your perception and just be.

I know you fear loss and change.

Impermanence can be daunting, but don't lose sight of what else it is. A giver of gifts and finder of paths.

One incredible aspect of the human mind is our ability to choose to change.

To acquire a skill.

To alter our thought patterns.

To adopt new behaviors.

These are all gifts of impermanence.

Our ignorance is impermanent. Our wounds are impermanent.

Those social forces we wish to alter or oppose are likewise impermanent.

If ever you feel you've stopped becoming and started simply continuing, remember this wonderful feature of your mind and your world.

Impermanence means possibility.

You may call upon this aspect of our reality, this doorway to intentional becoming. It is yours to seek, to adopt as an ally.

Think of the scope, grandeur, and mystery of the universe.

Think of the glorious feel of your own warm embrace.

Recall how your life arose from the singular, magnificent event we think of as nature.

A locust tree. The turning wheel of a galaxy. Nature.

Know that your mind and thoughts arose from that same nature.

These words and your understanding of them.

You are part of this tradition.

What does that tell you about your own possibilities?

Frogs, butterflies, jellyfish. Their lifecycles contain startling shifts in form and behavior.

And you?

Are you the same person you were five years ago? Fifteen?

Time and biology are agents of transformation, but so is knowledge. So is choice.

Leave room for the next version of you.

When we focus too much on safety, we focus on sameness.

Sameness did not bring you the things you love. Change did.

Remember that you are here to experience, to witness.

Life isn't a game you're trying to win.

Learn to take pleasure in playing it.

You might find yourself afraid to reach for comfort or healing or fulfillment. The act of reaching can feel like it opens you up to risk, to the bitter possibility of disappointment.

Here's something I have learned: effort will not always lead to the result you had in mind, but effort will not betray you.

Write the book.

Make the art.

Risk the love.

Strive for wellness.

Look beyond simple safety, simple continuance.

Effort does not make promises, but effort will not betray you.

Do you fear the risk of seeking to honor your intrinsic worth through care and consideration?

You might fall short. But even when we fail to find morels among the spring leaf litter, we are sure to find something we can choose to treasure.

Remember who and what you are.

Remember where you are from.

Fear?

You?

You who tasted oblivion before you knew your own name?

You who inhabit an ongoing explosion?

You whose each heartbeat becomes notes in the song of the universe?

You whose red blood is colored with iron built in ancient stars?

Friend, you have come so far already.

Please, take time to acknowledge the monsters you defy, the storms you weather.

There are those walking this earth to whom your life would seem an uninhabitable rock jutting from an angry sea.

Yet, here you are, stringing together days like flowers in your crown.

Respect your own vital splendor.

You are worth it.

You are worthy of peace and happiness.

You are not safe. You never have been.

From year to year, you have never been the same. The waves of change roll on and on, reshaping your coastlines, giving and taking in one continuous motion.

This version of you will not linger into the future.

You will, thankfully, be touched and altered by this life.

Yet, you are amazing, indelible, and part of an awe-inspiring family of natural wonders deserving of love, enthusiasm, attention, and effort.

Nothing can change that.

Your unique character and perspective, now and forever, is a part of existence.

Fear is natural, but couch your fear in the proper context.

Hold it up next to what you are, where you have been, and what you know.

See how your fear shrinks by comparison.

Fear does not control you.

Do not play dead for fear of the real thing.

Death does not need your fear nor your flattery through imitation.

Yes, we have limits, and change comes for our plans and stone monuments alike.

But change is what brought you here, and it is not your enemy.

Sometimes, we fear not how we might be altered but what we might miss. We fear the knowledge of all the things we will not have.

Other loves.

Other pleasures.

Other understanding.

Other lifetimes.

Things that call to us from beyond our own bounds and endpoints.

A fear of missing out.

But just as change is not our enemy, neither are our limitations.

Limitation gives form and shape to our lives.

It gives us purpose and character.

Poetry teaches us that limits enrich and nourish.

So too music, art, and cuisine.

Aesthetics and flavor.

Your favorite song does not use all notes simultaneously.

Your favorite art does not contain all colors and hues.

The sweetness in life is as much a question of omission as inclusion. Remember this when considering your own limits, when fearing the edges of your life and experience.

The answer to "what is enough?" is not "all of it."

Do not fear missing what is out of reach. The limit of your reach is a defining characteristic of who you are.

Fear has a purpose.

It is meant to keep us safe.

Yet, safe is so very far from the sweetest thing we may hope to be.

❖

September's insistence that it was still summer began to lose credibility.

The leaves were changing.

The air was changing.

In the height of summer, walking the woods, I often anticipated the coming autumn. Each time I had to pick spiderwebs from my beard, flick a deerfly from my wrist, or check my back in the mirror for ticks, I wished for the cold's return.

Beyond a farewell to biting insects and strong storms, I looked forward to being outdoors in my favorite season with a healthier mind and outlook.

The medication had helped.

CBT had helped.

I was gaining practice steering toward a steadier mood and using my newfound calm to establish a new, friendlier sense of day-to-day normality.

During those long, summer days, I began to feel I had all the tools I needed to push back against mental illness.

Like the summer weather, it felt simple and effortless. It felt like it would last forever.

I wouldn't have called myself depression free, but I was more than just coping. I was managing my illness, a thing I hadn't thought possible. I was rediscovering interest, pleasure, and enthusiasm. I wasn't chased through the minutes of my day by either shame or a death wish. I was becoming me again, the me I'd thought was lost.

The unforeseen downside of my hopeful summer arrived as the days shortened and the squirrels built new treetop shelters.

Fear.

I carried an unsettling sense that the season was not the only thing approaching a cyclical shift.

I mourned the opossum and my role in its death.

I reminded myself that some sadness is natural.

I reminded myself that I could recognize my influence without owning things beyond my control.

I reminded myself that every human makes mistakes.

These reminders felt simultaneously true and not terribly helpful. Some hurt is just hurt and won't be soothed by sense or philosophy.

I can make peace with that, but this particular hurt began to have a shadow bigger than its substance. My regret for the possum's death bled outward, spilled ink across the pages of my newly organized thoughts. The remorse was no longer just about the possum. The blot became a seed, became a sapling, became a thicket.

My hurt became, once again, hard to tie to specific thoughts, hard to grasp.

This wasn't about sunflower seeds and a dead possum. It was an old enemy moving in the shadows. It was winter catching up to me.

"I'm alright," I lied.

The summer was departing.

I started going for fewer walks.

I found excuses to avoid friends.

I quietly surrendered to a growing dread that my depression was not just an illness, but a planet at the center of my existence. A planet with its own unyielding gravity. And whatever honesty, kindness, or love I could mold into a healthier worldview—science and nature and magic and whimsy—would all eventually be pulled into its grim, implacable orbit.

I had come so far, and I'd believed I had made it past such thoughts, that I was beyond their reach forever.

This wasn't about the death of one possum.

This was me, after decades nearly drowning in storm-tossed seas, standing on dry land just long enough to realize that a tide beyond my control was surging back to swallow me again.

It's easy to be thoughtful and philosophical about the ocean when you are warm and dry. Infinitely less so when you are being swept out of sight of land.

Somewhere amid the fear, I heard Mark's voice from our first counseling session:

"Progress is not a straight line."

FALL

FALL IS CLOSING TIME in Ohio.

Lights out.

An empty street.

Staying up past our bedtimes in the big, too-quiet house of the world to read stories by flashlight.

Apples and pumpkins and nervous laughter in the neighboring cemetery.

The cold wind comes at night with a reminder of the importance of warm blood and warm rooms.

It whispers in the chimney like the rustle of roosting birds, telling secrets, saying, "Nature has a heart, and every heart must rest between beats."

Each pulse and breath and sentence and song requires stillness.

Requires endings.

Chapter 16

RACCOON

MY FAVORITE SEASON HAD ARRIVED and I was back in sweatpants, back to avoiding company, back to lurking in the dark kitchen and criticizing my reflection in the glass doors. Everywhere my gaze landed felt like a new kind of failure. The dark circles beneath my eyes. The muddy paw prints on our cheap, stick-on vinyl tiles. The columns of dirty dishes that smelled of sour milk and burnt coffee. Even the porch railing where I doomed an opossum with thoughtless enticements was beginning to lean, one more sign of decay, one more thing I was failing to hold together.

My depression had regrouped and redoubled its attack.

Admit your shame, it said. *Admitting it won't remove the knives from your guts, but come clean and maybe they'll stop twisting so much.*

It wasn't hard to capitulate.

I was ashamed. It was the old shame of a wayward life and mind. It was a new shame too. I had, so recently, felt like I had a handle on my mental health for the first time in decades. Now, that confidence seemed like foolish arrogance and self-deceit, just one more embarrassment. This new depression felt like a rebuke of the idea that long-term management of mental illness was even possible.

I stopped sleeping at night.

I woke up around 5 p.m. to see Leslie, then I would go to bed at sunrise.

Things felt easier in the dark.

I stood at my back fence at 2 a.m., watching the little lights flicker in the cemetery. I ate pretzels or string cheese or a sleeve of crackers and called it a meal. I played video games I didn't enjoy and listened to the same audiobooks over and over and over, unwilling to face the potential disappointment of anything new or challenging.

One night, in the hours just before dawn, I was sitting in the dark scrolling through Facebook, reading responses to a recent post I had made about my depression. My mind was busy disregarding the positive and supportive comments and focusing on the critical and dismissive remarks.

"Depression isn't real. Get stronger. Think your way through it."

"Have you had your testosterone tested? Maybe you're actually sick?"

"Exercise is the key."

"It's sounds like a spiritual problem. Is God in your life?"

"Whatever you do, don't poison your body with meds."

Atticus pressed his nose beneath the front door and growled. Atticus rarely growls.

My old paranoia flared to life, shouldering aside the fog of sadness. I grabbed a weapon and shrugged into my big, black coat before turning off our motion-activated floodlights and slipping out the backdoor. Perhaps it isn't logical, but I feel safer outside and I feel safer in the dark. I have uncommon night vision and a knack for moving quietly. I stepped into the tangle of a forsythia bush at the rear corner of our house and stood listening.

Silence.

Minutes passed.

The hushing sound of distant traffic. The metallic howl of a train whistle. A big dog's *woof* somewhere beyond the cemetery. The wind changed, carrying in a chemical smell like new plastic from the nearby industrial park. All around me, in every direction, people were sleeping while creatures at the periphery of human society moved through the shadows.

I quietly rounded to the front yard.

There was nothing.

I stood still and refocused on movement and sound.

I heard a faint rustle, then saw our big, blue plastic dumpster shudder.

I smiled, my tension vanishing with instant recognition.

"Ah. It's you," I said to nobody.

I moved closer.

After days of rain, our yard was a swampy mess and I saw the muddy raccoon prints on the blue plastic, confirming my suspicion even before I peeked inside.

I lifted the lid.

There in the yellow porchlight glow, a fuzzy, masked face looked up at me.

"You're upsetting my dog," I said.

The raccoon stared back. "What of it?" he seemed to say.

He had shredded the garbage bag and was picking out old pizza crusts.

"I'll check back later to see if you're trapped, tough guy."

It wouldn't have been the first time I needed to free a raccoon from a dumpster, but as I walked away the lid clattered and the raccoon climbed out and scurried off toward the shaggy evergreens across the street.

I chuckled and marveled at the lightness in my chest.

Even this small encounter with wildlife tripped-up my depressive thoughts, rerooting me in a sense of a larger world.

I went back inside, flopped down on the couch, and thought about disappointment, relapse, and raccoons.

❖

Nearly seven years earlier, just after graduate school, I had been feeling restless.

We both were.

Leslie and I were disenchanted with academia, and we wanted a dramatic move, something to carve a moat between the past we were escaping and the future we envisioned.

We wanted dramatic and we found it.

A single phone interview landed Leslie a marketing job at a small company in Tacoma, Washington. It came together suspiciously fast, but we decided to take a chance. We stuffed everything we owned into her old Toyota Matrix and drove 2,500 miles west.

After nearly thirty years in Ohio, the landscape of the Pacific Northwest was an illustration from a fantasy novel come to life. The ocean. The mountains. The towering fir trees.

Everything was on an unfamiliar scale. Crossing the mountains from the brown, flat expanses of Eastern Washington into the verdant costal region was like driving to another planet.

It was autumn, but the Seattle area didn't look like any autumn I had ever seen, all misty and deep green.

Leslie began her job and I spent great swaths of time sitting in a folding camp chair in an unfurnished apartment with a tremendous flea infestation. It was a ground-level one-bedroom that smelled like old tobacco and industrial cleaner, an off-white box in a large rental complex set into a wooded hillside. In the plus column, it had a working fireplace, something I had never encountered in any Ohio apartments.

We could have romantic fireside evenings with hundreds of our closest flea-friends.

After many weeks of job rejections, I attempted to find a volunteer position to occupy my mind, network, and get away from the hungry insects we couldn't seem to banish.

I had no luck.

Not with networking. Not with the fleas. Not even with volunteering. I had just earned a master's degree, and I couldn't even give my time away.

Each morning, Leslie drove off to a workplace of disrespect and sexual harassment, and I set out on foot, trying to at least be a moving target for despair. I had no transportation and lived in a blank box where tiny creatures were literally stealing my blood.

There was a manic electricity buzzing through me at that time in my life. I had stepped off the path to professorhood, one I'd been following for the better part of a decade. That path hadn't been working for me, but at least I knew how to follow it. Now I was in trackless wilderness and felt like anything, good or bad, could be rushing to meet me. I understood nothing. Not my future. Not the worth of my past efforts. Not even the landscape or geography around me.

We couldn't afford to live in Seattle proper, and I was both broke and frightened of navigating the bus system, so most days I set out toward the nearby woods. We lived in a semi-rural commuter enclave where thick forest was visible in every direction. There was a park a couple miles from our apartment where a wooden walkway over wetlands was shadowed by dense evergreens. It was a misty, mossy forest that was utterly different from the young, briar-haunted Ohio woods I knew.

This wilderness was giant and old, full of a beautiful, watchful silence.

A place of wonderful medicine.

I never saw an eastern gray squirrel or a bluejay.

I don't recall sugar maples.

No cardinals or goldenrod.

These were common sights in Ohio that, for the first time in my life, I was in a position to miss. There were, of course, many other worthy lifeforms. Douglas fir. Red cedar. Western sword fern and long-toed salamanders. And it was a treat to walk alongside so many new sights and sounds. They just didn't feel like home.

I wasn't in Washington long enough to develop a familiarity with the wildlife there. Still, the forest felt like a forest, a place where life became simple and essential and I could step away from anxiety and doubt and be present within myself.

A forest is a wonderful place to feel for the edges of your mind and body. To be both grounded in the moment and explore your sense of identity.

How ancient and powerful you are as the centipede crosses your boot.

An opportunity to be gentle.

How young and fleeting you are looking up at the treetops.

An opportunity to be small, to be humble.

Nature, in all its richness of form, has a way of enticing me to set my perspective aside and try on others. This is especially soothing when my perspective becomes a minefield of shame and regret.

My world in Washington was small and frustrating and confusing, but I was pulled to the trees most every day and, for a little while, they made me feel at peace with my uncomfortable present and uncertain future.

It's funny how often I realize my connection to nature and think I am realizing it in a new and final way, that this time the realization counts, that I've had the last word on the subject.

These are realizations about identity and, like much of the meaning in our lives, are never finished.

We are things of cycles and seasons.

We are repetitions and variations that build a layered pattern through time.

I knew nature was medicine when I was ten and when I was thirty. I know it now, in my forties, in new and different ways. The pattern grows and deepens.

For many years following, our stay in Washington felt like a wound. But that feeling also changed with time.

In the Ohio autumn of my present, as depression crept back into my life following the first truly successful treatment I had experienced, the important part of our time out West wasn't that it wounded, but that the wounds eventually healed. In Washington, I felt as lost as I could be, but that feeling had turned out not to be the enduring, defining moment I feared it was while living through it. Feeling lost is not evidence of a doomed future, not evidence we're fated never to be found.

My understanding of the stages of change began to inform my memories of that post–grad school year in Washington, adding nuance. I saw the backward-pointing arrows labeled "relapse" on Mark's stairsteps of change and understood how they were integral to the process. Our move across the country was an instructive setback, not a dead end or a final defeat. I thought of Mark's sensible assertion that every disappointment and backslide is simply data, not cause for judgement and guilt.

Failure was necessary.

Mistakes were inevitable.

Disappointment did not need to be damnation.

Our Seattle adventure certainly ended in disappointment.

It also ended with a raccoon.

❖

Leslie's job in Tacoma, which increasingly turned out to be tied to a petty, predatory, incompetent man, culminated in an unceremonious layoff. She overheard a conversation indicating that, despite knowing she had moved across the country for the job, her employer had always planned to lay her off once business slowed.

She was not terribly disappointed.

My freelance writing wouldn't pay our bills, so with our only real source of income gone, we made plans to return to Ohio. We gave away the one large piece of furniture we had acquired, a mattress, and shoved everything we owned back into the Toyota.

We had nothing, not even enough to pay the fee to break our lease. Considering my student loan debt, we had less than nothing. Leslie's parents sent us money to end our lease lawfully, though I was beyond caring about the potential consequences of fleeing the state without paying.

When it was time to go, I left before Leslie. Her family had booked her a flight from Seattle to Michigan to attend a cousin's wedding. I would be driving back alone.

The weather report predicted storms, and I worried that Snoqualmie Pass, an early obstacle on the eastward drive from Seattle, would close. So, I quitted our little apartment, and Leslie stayed one more night without me. She slept on the floor surrounded by her luggage, grateful at least that we had finally won the war with the fleas.

Just after midnight, she began to hear an odd tapping noise. She ignored it at first, then became aware of the subtle chiming of glass mixed with the taps and clicks. It was coming from the glass door that led to the rear of the apartment.

There was nothing back there but a ten-foot strip of wet, sickly grass and a wall of thorny blackberry bushes.

She was alone in a carpeted box with nothing but a suitcase and the scraps of an abandoned life. Something was trying to get in. Stealthy sounds at a door that led to a dark, soggy nowhere. A half-hearted border between our ground-level apartment and the woods just a stone's throw away.

She was unnerved, but there was nothing for it.

Skin tingling with gooseflesh, she clicked on the bare bulb that illuminated our tiny back patio and looked out.

An uncommonly large raccoon looked back, his tiny hands pressed flat to the glass, returning Leslie's gaze with placid unconcern.

It was an appropriately ridiculous and menacing end to our time in Washington.

She gave the raccoon a high-five through the glass, and returned to her spot on the floor.

❖

I made the drive back to Ohio in only two sessions.

I drove a sensible seven hours the first day.

On the second day, I decided I was entirely sick of the western United States and drove the thirty hours to my parents' house without stopping for anything but gas. No sleep. I don't even recall a bathroom break. I craved the Midwestern landscape with an intensity that propelled me onward long, long after I was not safe to drive. The gentle hills. The patchwork fields of corn and soy. Gray-skinned oaks and maples. Cardinals and goldenrod.

It wasn't a responsible choice, but I felt desperate to return to a place I knew. I had been a squirrel stuck on a roadway. A mouse trapped in a sink. A ratsnake poised on the edge of a busy highway full of crushing tires. I needed a way back to something familiar.

I was headed to my childhood nature and my childhood home. I couldn't stomach any more delay in getting there, not when there was nothing left keeping me away but my own physical limits, limits I gambled on ignoring.

My parents were kindly acting as a soft place for us to land and we were ready to take full advantage of their generosity. It was time to regroup. Returning to their place wasn't exactly what I had hoped for my future, but it was warm and safe and mercifully free of fleas, if not scheming raccoons.

We felt defeated. We had both worked many jobs. We'd struggled through college then graduate school. We'd earned master's degrees. We'd moved to a new city with high hopes and a job offer, and, a handful of months later, we were moving into my parents' basement.

Most of our lives had been spent on the path to something ambitious. No longer. My parents' basement didn't feel like a trailhead on the way to a career or a stepping stone leading toward a goal. It was a shelter from a storm, a place to hide.

We were on the defensive.

Still, we were fortunate and we knew it.

Few enough people have such warm, welcoming ways to fail.

It just wasn't clear to me when or how we would get back on our feet.

❖

Raccoons are opportunists.

They'll eat frogs or berries.

They'll forage for nuts or steal eggs from nests.

They'll dig through dumpsters or walk right in through your dog door and check your pantry.

Somehow, they look equally at home climbing a palm tree or snatching crayfish from a mountain stream or strolling down an urban alleyway.

I once growled at a raccoon that was approaching my campfire and he just stood on his hindfeet as if to say, "If you're feeling froggy, then jump big guy."

I laughed and shared my sandwich.

The primary thing raccoons seem to do is thrive.

They thrive in human spaces and wild spaces.

They are versatile, intelligent, and exude an air of crafty stubbornness.

It can be romantic to glean life lessons from herons and hawks and deer leaping through thickets, but there are many seasons in my life when what I really need is the teachings of raccoons.

The fall we retreated from Seattle was one such season.

The fall when my recent successes in fighting depression seemed to be slipping away was another.

I imagined these two disappointing autumns transposed, one on top of the other.

I saw the raccoon pawing at the apartment door in Washington.

I saw the raccoon peering out of my dumpster in Ohio.

Both creatures spoke with one mouth, asking an essential question, "Who says survival has to be pretty?"

I didn't want to hear it.

I had spent many months working with Mark toward a stable mind, and I felt like that work was slipping away.

I hurt.

It was the old hurt, winter's hurt, the kind of depression that sits on your chest until your ribs ache and your lungs burn, until you just want out.

My thoughts drifted back to Seattle, then to other disappointments.
My failed first marriage.

My ill-fated first attempt at college.

Lost friends and lost memories.

I thought of all the times I had taken a leap of faith and found myself
bruised, face-down with a mouthful of dirt as the reward for my efforts.

I was still attending counseling, though, and Mark reminded me that
the stages of change included relapse. That it was data, an integral part
of the process.

I could accept that intellectually. Emotionally, acceptance was harder.

I believed in counseling, but I also believed my pain. Counseling was
an hour a week. Pain was every day.

I had come so far. What was the point if it could all up and vanish
again with the changing of the seasons? How could I face my own hollow, devastated expression in the mirror?

CBT had taught me that dealing with such thoughts can be like fighting ghosts. They hit you, but you can't hit back. They aren't tangible
enough to attack while they float through your mind, ambushing and
vanishing like smoke and shadow, but with very solid fists.

The solution, for me, is to write the thoughts down.

So I took pen and paper and wrote down a question I had tuned into
through all the static inside my head: *what was the point of all this work
if depression can just return?*

I stared at the words on the page, a vicious idea given form and
trapped in place with ink and letters. It occurred to me that I could think
of past versions of myself who'd been haunted by similar questions.

I tried to write a few of those old questions on the page.

What was the point of falling in love if it can end in pain?

What was the point in trying to be authentic if it can still lead to rejection?

What was the point of all our work and study if we just ended up in my parents' basement?

All of those past questions were relevant and razor sharp … once upon a time.

But no longer.

With time and distance, they had become inert, maybe even a little quaint and nostalgic.

Going through a divorce hadn't ruined love for me forever.

I lost no sleep over friendships that ended in high school or embarrassments from twenty-five years ago.

I didn't regret my education just because it didn't lead straight to wealth and success, nor did I regret our attempt to move to Washington even though it ended in retreat.

My mind is a richer place for these efforts and experiences.

Data.

I had tried and lost before. I would try and lose again. Why should my depression relapse be any different?

I could turn away from my vulnerability and potential disappointment by letting my fear and ego steer, but I knew where that toxic mindset led. Playing possum doesn't protect, and arrogance isn't strength.

Failure hadn't broken me in the past, and it wouldn't break me now.

I underlined the *my parents' basement* question, recalling the details of our short time out West.

I remembered the fleas and how their bites left lines of red dots on our ankles that made Leslie cry in the shower as she got ready for work. I could hear her softly sobbing over the sound of running water.

I remembered trying to cook Thanksgiving dinner and laying our little feast out on the floor because we didn't have a table.

I remembered how I was drunk from sleep deprivation when I finally crossed back into my little hometown after driving for thirty hours straight.

Those weren't happy memories, but I didn't want to erase them. I had erased too many already. The memories weren't bleeding wounds. Just faded scars. Tokens of life lived.

I recalled unpacking the Toyota. Moving our things into my parents' basement, it was unclear how we would ever recover and set out on our own again.

It remained unclear until ... one day ... it was clear.

It was complicated until it wasn't.

Stubborn hope and steady effort feel slow and daunting at the time, then simple in retrospect.

There, in my parents' basement, I wrote ten template resumes emphasizing different skillsets and experiences. I applied for jobs everywhere.

Eventually, a kind fundraising manager named Joe Neal took a chance on me and my insistence that I could learn to manage donor database software. I was hired at Franklin Park Conservatory and Botanical Gardens in Columbus, Ohio, a half hour from my parents' house. Suddenly, I had a foothold in a career outside academia. I learned contact management software. I gained experience in fundraising and nonprofit work. I did data entry and helped implement a new point of sale system. I took on writing and marketing projects when I could and gathered new responsibilities. I shifted career paths.

Soon after I was hired, Leslie found a job, and then we moved into our own place near Columbus. We were knocked down and, with generous help, we got up again. We schemed and scavenged and, eventually, got what we needed.

Mark strolled into my memories of those hard times and the recoveries that followed. He said one word.

"Data."

Meanwhile, something else, something with furry little hands, tapped at the backdoor of my thoughts.

Sometimes I need a heron to be my contemplative guide. Sometimes I need a deer, antler-crowned, or a bluebird like a living shard of cloudless summer sky. But that autumn when my newfound mental health was collapsing around me, bedraggled in sweatpants, hounded by barking worries like hunting dogs, I needed a raccoon, a trash-goblin, a clever little cutthroat with Dorito dust on his whiskers to stand among my dirty dishes and whisper, "It's not what you eat or how you eat, but *that* you eat."

Yes, nature is soaring eagles, hundred-foot pines, and leaping marlins glistening above the seafoam, but it's also raccoons armpit deep in a discarded peanut butter jar.

My raccoon tapped on the door to my thoughts, and I decided to let him in. As he rummaged through my snack drawer, he said, "Stop thinking survival needs to be graceful." He munched Goldfish crackers and spit crumbs while giving advice.

"Who cares if you're a mess? You think you're in a beauty contest right now? Trust me, you ain't. You think every day needs to be a poem? Pal, you're mentally ill. Cheat. Steal. Eat fast food twice a day. Whatever it takes. Just breathing counts as 'doing it.' This is what a win looks like right now, got it? If a stray dog chases you up a tree, better to be grateful for the tree than pissed that dogs exist."

He shrugged furry shoulders and went on eating.

I glanced down at my paper. I had a list of rhetorical questions, old and new, and an answer in the form of a raccoon doodle. I scribbled out the questions and circled the raccoon.

The effort of just enduring deserves to be celebrated. Actually and vehemently celebrated.

Enduring isn't a consolation prize. It's a win. Change is always coming. Enduring is sometimes our way of meeting that change halfway. In some contexts, in the midst of some hardships, surviving counts as thriving. That was for me to decide.

Success is always mine to define.

Our Washington misadventure was many years ago. The fleas. The rejection. The move back. All old hurts without much bite left in them. One day, I knew, the brutal disappointment of realizing my depression was not permanently banished after my successes with medication and counseling would also fade to a pale scar. I just had to endure, to keep stacking up the days and making space for new possibilities.

I would take my meds.

I would walk when I could.

I would go to counseling.

I would be disappointed and an absolute wreck and call that good.

My winter, spring, and summer struggles were not for nothing, and the lessons I'd learned were not a waste.

I had more tools than I did a year prior.

This was the process, collecting moments, gathering understanding, welcoming vulnerability and experimentation, and letting time do what time does: bring change and perspective.

Chapter 17

SHAGBARK HICKORY

CERTAIN TREES WAIT FOR ME.

There's a shagbark hickory on the "primitive trail" at Gallant Woods Park that waits for me. It's a ten-minute drive north of my house, past baseball fields and new subdivisions with big, uniform houses and tiny yards.

It's a friend and a landmark.

That hickory has been standing there throughout my life. It was there when ten-year-old me sat high in the branches of my sugar maple. It was there when we ate our overcooked Thanksgiving turkey on the floor of a tiny apartment near Seattle. It was there when I saw a heron and felt the winter wind chilling my tear-streaked face.

In the spring of my healing, when I was reawakening to my love of nature, I finally noticed the tree, tall and broad, jagged strips of thick bark hanging away from the trunk like matted fur.

I didn't know its species name, and that ignorance became a lesson in perspective. At first, I scolded myself for forgetting the name of such a common, familiar, recognizable tree of my childhood. But as I learned to be kinder to myself, I began to think of shagbark hickories as reminders that many worthy things hide in plain sight, within our simple, natural ignorance.

My learning journey with hickories is not an embarrassment. It's a treasure.

I knew hickories, then forgot them. I knew this particular tree by sight, then feel, then name. I knew it better still as I watched it change through four seasons. Better still when I knew the sound of squirrels racing up the tattered bark. Better still when I walked among the green husks of hickory nuts.

What I learned coaxed me to learn more.

I learned that some bat species shelter in the spaces beneath the shaggy bark.

I learned that shagbark hickory nuts are one of the few native nuts that are good to eat raw. They are loved by many species, from humans to mice to black bears.

I no longer regret that I did not know these things.

I am only grateful I had the opportunity to learn them and that I have barely scratched the surface of all there is to know. Part of me is quite sure I once knew some of these things and forgot. Part of me is quite sure that I will forget again.

This, too, feels perfectly natural.

❖

In the winter of my pain, I was frozen, depression driving me toward sleep and death, while love pricked me onward to find shelter.

Spring arrived, a trickle of water and a hint of green, a frightening call toward hope and growth.

In summer, long, slow days of bright clarity, big thoughts, and ambition for better futures.

These were three seasons of growth. I grew in awareness, health, and knowledge. They felt like steady, if not quite linear, progress and they seemed to promise more progress to come.

Yet autumn followed, and it had something else to say.

It said, "Seasons aren't just for hickory trees."

It said, "Not all progress involves conspicuous growth."

My depression is never fully defeated.

That's okay.

It must be okay.

It is beyond my control.

And I know in my bones that "Does it last forever?" is a poor question to ask any of the good we encounter in life.

I don't think this means I need to be cheerful about my relapses. Depression hurts. It hurts a lot. It is exhausting and seems to attack not just how I feel, but who I am.

That second part, the identity part, is all about shame, one of depression's chief defenses against treatment.

When I faced a sudden return of acute emotional pain that fall, one of the first ways I realized that all my effort and growth over the last year couldn't be washed away by a backward slip was through my new relationship with shame. It was different this time, different because I didn't just accept what my illness had to tell me.

When I hurt and struggled to get out of bed as October rose from the scattered remnants of summer, shame whispered, "See, you are broken and pathetic," but I had answers this time that I didn't have before.

"No. I've seen you respond to treatment. You're an illness. A headache. A runny nose. I have weathered you before and I will again. You're just a thought. Not an action. Not an identity. Not who I am. You're just a thought, and I can build thoughts of my own to counter you."

I recalled my first conversation with Mark about the stages of change, when I asked him what it means to honor the messy, nonlinear process of altering ourselves.

"It means that we don't expect to be perfect. It means than when we miss a goal and slip backward, we smile and acknowledge that it isn't a tragedy, that it is the process working as we expected. When we give ourselves that honest, essential generosity, it isn't so painful to try for that next step because missing isn't a refutation of what we have accomplished or who we are."

I was still seeing Mark. That continuity of care was another new tool in my collection. Shame was a frequent theme in our talks, and I often returned to our stages-of-change discussion.

One autumn afternoon I asked, "But how can moving backward still count as progress?"

He just nodded at me and raised his eyebrows.

I already knew the answer and he wanted me to say it.

"It counts as progress because it is progress."

"There it is," he said.

When my depression returned that fall, it felt like the same depression, but it wasn't the same. It couldn't be. It had changed because I had changed.

I was changed by the efforts of Leslie, friends, and loved ones. I was changed by abandoning detachment and reconnecting with nature. I was changed by seeking and receiving treatment.

I was not the same person I'd been in January. I was a person who had new rebuttals for shame. I was a person who had risked healing and experienced the rewards. I knew such things were possible. And, now, I felt a growing confidence, despite my sadness, that I would soon be a person who faced an inevitable, natural relapse and, with effort and intention, arrive on the other side wiser and stronger.

I could be disappointed and frustrated without feeling broken or worthless or beyond hope.

Sadness, even hopelessness, did not need to be joined with a surrender to despair or unchallenged shame.

In many plant species, prompted by the cold and shorter days, fall is one of the most active times for root growth. Plants tend to lose less moisture in cooler weather, and energy no longer directed toward producing shoots and leaves may be redirected to growth beneath the soil.

Not all growth is obvious or contingent on gentle, summer days. A bad day or a hard season doesn't equal a bad person or a hard life.

Days and seasons change. That's what they do. They are temporary.

Challenging times can prompt new kinds of growth. We, too, can build deeper, stronger roots.

I was far from happy that October, but I found that I was anchored in new and sustaining ways. I was becoming rooted in a new understanding of myself and my illness. I had possibilities. I had experience. I had new tools. Now, I had the chance to test what I had learned against the coming cold.

❖

In late October, I had one of the worst depression days I can recall.

I'd stopped working at the wildlife center, having eventually discovered that event planning and wedding-venue rentals weren't a good fit for me. I planned to return as a volunteer, but hadn't yet. I'd unexpectedly begun making enough money from writing and podcasting that I was starting to turn away from the idea of conventional employment altogether.

Once again, nobody was counting on me to do anything at all, at least not in a way that required a rigid, daily schedule.

On this day, after a dozen false starts trying to get out of bed, I stood at the kitchen sink and ate dry, sugary cereal from a box, feeling like I was going to vomit from pure despair.

Hungry for distraction, I tried to read the ingredients, but I couldn't focus my eyes on the words.

So, I just stood and mechanically chewed, trying and failing to silence my thoughts.

Every wakeful moment was misery.

I had slept ten hours and was very ready to go back to bed.

My joints hurt, especially a few troublesome fingers. Old jiu-jitsu injuries reminding me that they never quite disappear entirely.

I went to the bathroom and noted from my dark urine that I was very dehydrated. I would need to drink water eventually.

I stood at the mirror, studying my face.

My beard was going to seed, new growth up my cheeks and down my throat toward my collar, changing my look from a choice to a consequence.

I had made plans for the day: work on revising my soon-to-be-published poetry collection, then outline a new podcast episode.

I looked at my reflection and knew none of those things were going to happen.

Old internal monologues popped into my head, endless recriminations for the sad, tired man in the mirror. Weird. Pathetic. Useless. Broken beyond repair.

I swallowed and calmly labeled those thoughts in crisp, black marker. "Symptoms."

I set them on a shelf in my mind, picturing it as an old naturalist's curio cabinet, unwanted thoughts alongside feathers and skulls and bits of moss.

I looked myself in the eye, studying my blue-green irises, and tried something new. I tried sympathy.

"I'm sorry you're feeling these things," I said in a horse whisper.

The tears came.

"You didn't choose these feelings. You didn't do anything to deserve them. They won't last forever."

More tears.

The words helped, but I wanted to die all the same.

Suicidal thoughts had returned.

They could go into the curio cabinet too.

I sighed and tried to think about my day.

My eyes drifted to the cabinet.

The death wish was real. There it was squirming in its mason jar next to a pink hunk of granite.

I knew I wasn't going to act on it, but it was real all the same.

Maybe I couldn't banish depression or will myself to have a productive writing day, but I had other powers.

I could change my goals to fit my circumstance.

I could change what "success" meant for that day.

I could go outside.

My friend the shagbark hickory was standing out there and maybe could use a visit. I certainly could.

The idea of leaving the house flooded me with redoubled exhaustion. Bed was so soft, and nobody was there to stop me from climbing under the blankets.

I pushed back.

If you're brave enough to die, you're brave enough to endure fatigue and take a drive. You're brave enough to go and see the hickory.

Leslie was at work, but I tried to picture her face and hear her encouragement.

The dealmaking began.

"Okay, but I don't have to shower."

Fine.

"Okay, but I'm going to eat something unhealthy on the way."

Fine.

"Okay, but it isn't going to help."

We'll see.

I begrudgingly admitted to myself that it could hardly hurt any more than staying in my silent house, alone with my pain.

So, slowly, steadily, I pulled on my dirty jeans and boots.

❖

It was a chilly, wet 11 a.m. on a weekday in October.

There was nobody at the park.

Why would there be? Other people have real jobs.

Depression always has a comment.

I didn't want to be there, but I was committed to running a small experiment. I wanted to see if it was possible to have a miserable, successful day. Whether I like it or not, all my versions of success require effort. It's one of my old beliefs, and I can bend it, but not break it. Human worth is innate. Success means overcoming resistance. It means work.

The air was crisp and misty.

I trudged through squelching mud, churning red and gold leaves down into the hungry earth. On that walk, I didn't feel happy with either my beliefs about success or the day's experiment, but I asked my legs to move and they did. One more commonplace miracle.

Moment by moment, stubborn step by stubborn step, I reached the shagbark hickory.

I reached out to the rough bark.

The cold, wet, depressed and contrary part of my mind wanted to be cynical about the tree and dismissive of its presence. I wanted to mock myself as a cliché, a crunchy new age fake. I wanted to call the day a waste.

But that part of me fell silent when my fingers met the tree.

Here was eighty feet of hickory, over a century of life, turning dirt and pale Ohio sunlight into food I could eat straight from the ground, into the oxygen I'd been breathing since before I knew what breathing was.

There was no ignoring it.

I could feel it there.

Really feel it.

Alive.

Electric beneath my aching fingers.

An ancient creature, silent and grand, beside a crooked little footpath.

Family.

That hickory stood tall over me and, for a moment, it was the only tree in the woods.

The only tree in the world.

The first tree. The last tree.

The concept of a tree.

I tried to calculate its value in that moment and saw how absurdly useless the word *value* is when applied to nature, like trying to pin a price tag on rain or sunlight or gravity.

I tried to imagine the tree as human technology, and suddenly I was standing in the presence of an impossible work of science fiction, a solar powered, self-replicating, carbon-fixing automaton producing oxygen and building materials on a planetary scale.

I tried to shift genres, fantasy and magic, and found I was a small, fleeting pilgrim who walked in the shadow of giants, living off of their sighs.

How much of my own form and evolution was quietly shaped by trees?

An apposable thumb for climbing.

Shelter and refuge from predators.

Fuel for cooking, unlocking more nutrients to build larger brains.

I know that touching a tree does not cure depression.

Of course I know that.

I also know that, on that October day, the hickory didn't need to cure my depression in order to take away its power.

That tree connected me with all trees, with our big world and our long, shared history. It brought me outside on a day when my inner life had become small and smothering. The bark of that tree was wet and cold and the touch of it still filled me with wonder and warmth.

My realizations about the tree, the fundamental aspects of its form and history, the meaning we made together, would not be brushed aside by gray weather or one season's depression. Those thousands of pounds of living wood, soaring branches and deep-delving roots, aren't defined by any one challenge or hardship, not by an illness or an injury or countless other vicissitudes of time and chance.

The roots it grew through a hundred autumns have lasted the most bitter of winters.

❖

There's a Nordic saying about the weather that I dearly love: "There is no bad weather, only bad clothes."

The shame of my mental illness often hits hardest when I find myself falling short of my own expectations.

The thing is, in the tension between ability, illness, and expectations, it is clear which element I can most easily manipulate: my expectations.

I cannot control the weather, but I can dress for it.

I cannot always control my depression, but I can control the ways I choose to mitigate and contextualize it.

In this sense, I like to think of the ups and downs of my mental illness as "brain weather."

If I plan a picnic and I need to abandon that plan because of a sudden thunderstorm, I may be disappointed, but I wouldn't be ashamed.

The arrival of that thunderstorm was not my fault. I did not cause or invite it. The same is true for mental illness. Why feel shame because of challenges beyond our control?

The brain-weather metaphor is helpful in another potent way.

Weather, good or bad, is not malicious.

That thunderstorm that canceled my picnic is not my enemy.

The thunder and lightning didn't spring from a grudge against me personally. Weather is a natural phenomenon and does not have intent. My illness is no different.

I didn't choose depression.

I didn't summon it.

It doesn't hate me and it isn't willfully harming me.

It is a known, widely observed, natural phenomenon.

It's weather.

Natural hardships like storms or depression aren't punishments, and they don't speak to our merit or shortcomings. They are neither justice nor injustice.

We don't own or invite them.

They aren't evidence of anything.

We can sharpen our forecasts and acquire more hard-weather gear, but there is no blame to be assigned or rebuke to be deciphered in our struggles to stay warm and dry.

My depression, like weather, is a natural thing, and natural things follow seasons and cycles.

Learning this was, itself, a new way for me to dress for my inclement mental weather. A new raincoat hanging by the door.

Chapter 18

LITTLE BROWN BAT

I'VE ALWAYS HAD A REPUTATION as a good person to call when a wild creature is out of place. Once, as an undergraduate at Ohio University, I was invited to a different floor in my dormitory building to remove a bat. I can recall a half dozen other times I've needed to remove a little brown bat from an indoor space.

Flying above carpeting, winging over lamps and furniture, they look so alien it feels like an elaborate prank. Based on the reactions I've witnessed, many folks think of an encounter with an indoor bat as something akin to a Halloween decoration springing to life in order to menace them. I adore bats, yet, even for me, meeting one next to a sofa is startling.

I don't regularly see bats up close. When I do, it's a special occasion. They are beautiful, elusive creatures of the twilight. And they remain so until your wife wakes you up because one has crawled in through the window air conditioner and is clicking and swooping just above your heads as you sleep.

That fall, we had such a visitor, not for the first or last time.

I was sound asleep when Leslie touched my shoulder.

"Jarod. Psst. Jarod, I think we have a bat."

I opened my eyes and took time processing the words.

"Ok," I said. "Are you sure?"

"Listen."

I did.

A chirp. A pop. A click, moving from one side of the room to the other. No doubt. We had a bat.

I yawned.

Two opposite thoughts wrestled for control. Bats are exciting. I am in the middle of very uncooperative brain weather, and all I want to do is sleep forever. If I'd been alone, I would have gone back to sleep, thoughts of rabies be damned, but I wasn't alone.

Your love is asking for your help. Dress for the weather and get moving.

"Yep. I got it. Just a sec," I said.

"I can get it if you want, just thought you should know before I turn the light on."

"Team effort?"

"Team effort."

We turned on the lamp and saw our guest looping back and forth across the bedroom.

Thankfully, we hadn't let the dogs sleep with us that night.

I have a sophisticated bat-catching strategy involving a T-shirt and patience.

I wait until the bat tires and lands, then I net it with whatever soft, lightweight shirt I have available. After that, I carefully bundle the fragile creature into the fabric and carry it outside.

Leslie and I each grabbed a shirt and sat on the bed, watching the bat, our faces swinging back and forth as if taking in a nocturnal tennis match.

Once, the animal seemed to be on a collision course with my forehead, but he clicked and swerved away.

It occurred to me that, for a moment, my face was translated through echolocation into a 3D map in a bat's brain.

There are so many unimaginable ways to know this world, and my way is only one, thin cross section of the whole.

Eventually, the little visitor landed on the curtains closest to me, and I nabbed him with an old shirt from Ohio State University's Newark campus.

"Got him."

I shifted the fabric until just the bat's upturned face was free. The little winged mammal opened his mouth showing off perfect rows of sharp, white teeth.

"Don't worry. Just helping you get back outside," I said.

The threatening jaws stayed open.

I delivered the bat to the back porch railing, where an opossum once stood, and opened the shirt, careful to keep my fingers away from those tiny, gleaming fangs.

He got his bearings and fluttered off into the night, vanishing.

I tossed the shirt in the laundry and went back to bed.

Leslie was already asleep.

❖

There are, of course, plenty of sensible reasons to exclude certain animals from our homes.

Contagions and parasites compel me to avoid sharing my intimate spaces with mice or bats or raccoons.

Likewise, I stay on the lookout for rare but dangerous creatures like black widows, brown recluses, and copperheads.

Yet, increasingly, when I carry a harmless spider or a stinkbug outside, I am doing it for their benefit rather than mine.

When a praying mantis or a tree frog takes advantage of the insect-attracting properties of our porchlight, I don't shoo them. I name them.

My depression is isolating. Many elements of modern living are likewise isolating. I sense a strange, pervasive pull in the air to stay inside, shop online, and discuss nature as a thing we visit when time allows, a thing that is "out there."

Touching a tree in a pristine forest counts as communing with nature.

What about touching that bit of lichen on the sidewalk?

What about touching my own cheek?

As someone who worries about isolation and who easily falls into the trap of all-or-nothing purity thinking, I worry about our delineations and binaries. It's the same purity trap that condemns us for not advancing all worthy causes equally, or for failing to be our hypothetical "best" selves. These are rigid, imaginary ideals that always clash with real life and the limits of real people.

It can feel like a violation when we spot something wild crawling or flying across our tidy living spaces, but I think the root of that feeling is the attack it represents on our illusion of separation, seeing nature in "nonnatural" spaces.

The reverse can also be shocking: when we encounter human influence in wild spaces. My friend Rebecca Helm, a marine biologist, recently told me about creatures who make their home on human trash floating in the oceans and how it took a mental shift for her to imagine floating plastics as being part of a functional ecosystem. Human development has vastly reduced the amount of driftwood that once reached the seas. Now, creatures that historically relied on that driftwood are adjusting to the newly available human-made floating objects. These creatures never got the message that human things aren't allowed to count as nature.

Again, health concerns are not an illusion. Many kinds of pollution are unquestionably damaging, and some creatures pose serious health risks in our homes.

Please don't leave your backdoor open and host raccoon parties.

They *will* accept your invitation.

That said, this idea that humans and human spaces are clean, sophisticated, superior and fully outside nature is ridiculous and damaging.

We are nature.

Our bodies are walking ecosystems.

The average human body has as many bacterial cells as human cells, all of which, human and nonhuman, through their oxygen, water, and nutrient usage, are in constant conversation with the wider world beyond our skin and awareness.

What are the indoors in that context?

What does visiting nature mean in that context?

Our viewpoints on this subject have impact.

I hear people talking about climate change as if it's a problem that is safely outdoors, as if a door is a magic talisman that removes us from the world that sustains us. We are neither gods nor alien visitors to this place. Our front doors are not gateways to another dimension.

Our bodies have never stopped being connected with nature, but our imaginations have. Culturally and psychologically, many of us need to reconnect our understanding, our identities, and our personal meanings with nature.

This isn't just important because it reflects an undeniable truth.

It is important because this truth, given its proper respect and consideration, is kind and nourishing. It's a warding spell against isolation.

It's a path to curiosity and enthusiasm, to enriching our lives through enriching our worldviews.

We often feel trapped inside our lives and homes and workplaces— but look around you. Your shelter may have walls, but your home is a vast wilderness you share with 9 million other species. That wilderness is not just "out there." It's in you.

I'd argue that those nonhuman microbial cells within your gut are still you. The water passing through your body is still you. You are a crowd within a crowd within a dynamic sea of staggering scope and diversity.

Walls may block our vision, but not our essence.

Try to really feel, for a moment, what you are. Not who. What.

Your blood is mostly water.

It uses star-forged iron to bind oxygen that was built by living things on Earth using the energy of our sun.

Water. Earth. Air. Sunlight. Life.

Plants and animals.

Unicellular and multicellular.

Elements and compounds.

You did not spring forth whole into this world, nor are you a matter of pure chance. You are a mix of very human choices and kindnesses, paired with universal forces older than the first star.

You are a grand collaboration.

Your life is not one thing.

You may feel separated from the natural world, but just look at the family resemblance.

The DNA we share with other lifeforms.

The way Earth's gravity is written into our bones, our circulatory systems.

You were not just born to this place.

You were born of this place.

Even your mind, the seat of your consciousness, is as natural as any green, growing woodland.

It arose in the world in the same way as a finch's wing or a cricket's song.

An orca prowling a kelp forest.

A luna moth, green as sea glass, fluttering in the branches of a sweet gum tree.

You, just as you are here and now, are nature.

Wherever you are, the part of you that's awake and reading this is deep in nature. Doors and walls have no power to change that. You may be sitting in a comfortable chair beneath electric lights, but your thoughts echo from an ancient wilderness.

How truly strange.

Sometimes I think there is nothing any of us could have done to earn a punishment so severe or an award so grand as being alive, but, of course, nature doesn't work that way. Nature doesn't measure deservedness. Nature is not fueled by human meaning. We are.

All of us here, drawing breath, are united in this shared circumstance of profound, exquisite strangeness.

We are ourselves.

We are each other.

We are the landscape.

And, within and without, the mystery of this experience continues to unfold.

I cherish a certainty that there are secrets beneath fallen logs, floating in the bright sky, and within my own skin that will not be known in my lifetime. I likewise cherish a certainty that these secrets are worth exploring. I hear these almost-truths chiming like distant music, and I'll walk toward that song with simple gratitude as long as I'm able.

Yes, there is suffering in this world.

And cruelty. And unwelcome endings.

We tend to notice the cruelty of our world while overlooking the kindnesses.

Why must nature contain pain and sadness? Why fear and loss and predation?

I understand why we ask such questions, but if we're going to play with hypotheticals, why should the Earth feed us or protect us?

Why should soil and rain and sunlight provide for our needs?

Why should the ocean and the forests supply our oxygen?

Why should our atmosphere shield us from cosmic radiation?

Why should gravity tether us to a warming, nourishing star?

Why should we assume these kindnesses of nature are defaults and take them for granted?

We can find reasons to hate our world, shortcomings that we would wish away if we only could.

We can also find endless reasons for gratitude.

The choice is ours.

My depression means I often have an uphill climb to reach gratitude, but nature helps me get there. I reframe basic facts. I make an active project of willfully noticing kindness and magic. I reject artificial boundaries between me and my world. I acknowledge the tolls of living without ignoring its gifts.

When I encounter reminders of a wider, wild world in my home, I welcome them when I can.

A lost bat, perplexed by the geography of a bedroom, is not an ill omen or an attack from an insurgent force. It's a rare chance to see a hidden wonder up close.

A spiderweb scribbled beneath my basement stairs is not a threat from an invader. It's a note from a distant cousin, written in a language older than the dinosaurs.

Please, enrich your life and our shared world by seeing nature for what it is.

Wonder. Kindness. Family.

Not perfect, something better and realer than that. Worthy of your care and attention.

Nature and our connection to it contains all the elements to construct the meaning you need, that humanity needs, but the decision to build is yours.

It's easy to look at the contours of a forest or the glittering arm of a galaxy and feel a bone-deep love for nature. It is less easy to remember that the contours of your own body represent that exact same nature. You are as worthy as any postcard landscape and deserve the same love.

Chapter 19

TURKEY VULTURE

I HOPE DEATH DOES NOT INVADE your thoughts as often as it invades mine.

I hope that you do not suffer from depression.

I hope you cannot relate to much of the pain I describe in these pages.

Even so, you probably have a memory of being home sick from school with a fever and an upset stomach. Do you remember the chills and strange dreams? Do you remember a feeling of betrayal from a body you thought you knew well? That you thought you could trust? Do you remember how the illness made you feel trapped within yourself in a new and disquieting way? How you noticed little things suddenly magnified by your discomfort? The texture of the couch. The individual fibers of your blanket. The way certain colors or sounds which had never concerned you before now seemed jarring, and out of place.

When something in me feels wrong, the details of life gain new importance. The balance of my focus shifts in unpredictable ways.

I think many of us have had that experience, one way or another.

My depression, when the numbness subsides and I'm able to reject wholesale despair, can lend strange colors and contrasts to my worldview.

Very often, my attention is drawn toward endings.

I may not have a choice about the frequency with which I think of death.

But I can choose how I think about it.

❖

November was wet and unseasonably warm.

I walked in frayed woodlands and skirted still ponds that impassively watched the heavy Ohio clouds with blank, liquid stares.

The air that was so recently summer nectar and birdsong was now stagnation and decay.

Winter was on its way, and much of life got out of its path.

But not all life.

Walking at Gallant Woods, I passed a rotting log and stooped close to see the moss and fungi growing on the dead wood.

Nearby, the sound of a foraging robin tossing leaves aside drew my attention.

A gray, wet November is like depression. Its broad impression seems uniformly bleak and dreary, but when interrogated for more nuance, unexpected contrasts can be found among the drifts of leaves and gnarled roots.

Suddenly, a patch of moss is a striking splash of color.

A robin rustling the leaf litter is a dramatic din.

The pain of depression is pain and November gray is gray, but I was learning that neither nature nor my mind was ever just one thing. There was relief hiding in the details when I could bring myself to look.

A dark, distant shape, moving in the corner of my eye, glided toward the ground. I turned and went to see what had landed. It was a turkey vulture, alighting next to what was left of a week-old deer carcass by the roadside.

One of my favorite birds. The turkey vulture. *Cathartes aura.*

A scientific name that means "purifying breeze"—apt for creatures that actively remove pathogens from the environment.

"Purifying," but not the smothering, abstract purity of all-or-nothing thinking. Ground-level purity. Small, tangible acts that benefit the animals sharing the vulture's habitat. Vultures don't eradicate all disease. They improve their world within the bounds of their own needs and limits. Not a deluge. Not a hurricane. A breeze.

I stood still and watched.

Turkey vultures tend to be shy, skittish birds, so I kept my distance.

She walked and hopped, spreading her wings and tilting her pink head from side to side, looking for threats. She paused and stood tall over the remains, wings outstretched, then folded, a moment of stillness and watchful silence.

She bent forward and began her work.

Leaning against a young hickory, I took out my little notebook and wrote a short poem:

Vultures are holy creatures.
Tending the dead.
Bowing low.
Bared head.
Whispers to cold flesh,
"Your old name is not your king.
I rename you 'Everything.'"

That deer was still mostly shaped like a deer, but not for much longer.

It was the afterimage of a deer, like the spectral silhouette of a light-bulb lingering in the dark after the switch is flipped.

Soon, shepherded on by the vulture and other agents of change, the deer's body would become the landscape that had once fed and sheltered the living animal. The grass nourished by the remains would, perhaps, feed another deer that might, one day, meet this same vulture's kin near this very spot.

The deer would rejoin the world that made all deer, and vultures, and wandering poets leaning on trees. It would be renamed "everything."

I watched the vulture having her meal and asked myself a familiar question.

Is this sad?

My natural, easy answer was, *no.*

The poem I had just written said, *no.*

But it's not quite that simple and I knew it.

I didn't know or love this particular deer as I know and love my family or myself.

Was I going to stand there and pretend I could view all death as one uniform monolith?

A death is a death?

No.

Not quite.

That view would be a disservice to the acts of personal meaning-making I cherish.

The concept of death and the death of a loved one are two distinct things.

I don't wish to be aloof or dismissive about death. That's too expensive a solution to my constant thoughts of mortality, sheltering me from sadness at the cost of sincere emotion. Life has weight and meaning and so too does its end.

Individuality is real. The diversity of self and perspective we celebrate is real. Why else would you be reading about another person's thoughts? Why would I have bothered to write them down or share them? The loss of a unique creature, a unique perspective, is real and meaningful.

Loss is loss, and I don't want to pretend it's not in service to some broad, sweeping philosophy of acceptance.

We all know loss, but death carries special weight that can bend and strain our worldviews. For many of us, me included, knowledge of our death, of death in general, its certainty and finality, can make this life seem purposeless or downright mean. Death can make nature feel unkind, like a rigged game, an insulting pantomime of existence masquerading as actual choice and agency.

That's the part I want to challenge. Not the idea that death has weight and meaning, but the idea that it's a menacing aspect of a cruel reality.

Death is not evidence of an unkind or unjust existence. This can be true even while we honor the natural grief we feel surrounding loss.

Let's think about loss, and let's start by bringing it home to the self.

We become different people over the course of our lifetimes, our outlooks and personalities, our cells and our connections, but we don't erect headstones for past selves. Even so, we know loss of self as intimately as we know our earliest memories.

Still, biological death seems different.

It is hard, as a living, thinking being, to view our end as a part of natural, wholesome cycles. So much of our biological hardware and software is focused on preserving the self and furthering our species.

From our limited, subjective viewpoints, death can appear to be just absence, a loss without renewal, perhaps even a theft. Certainly, if we focus on maintaining self and safety as our greatest goals, then death is the ultimate threat to all we hold dear.

But that outlook doesn't feel quite right to me.

A calm, poetic part of me sees death as something akin to the water cycle.

Each human life is a raindrop falling toward the sea.

The individual drop loses its shape when it reaches the waves, but what has been diminished? What has been lost? What exactly do we mourn?

Of course, I love some of those individual water drops. I love their exact shape and composition. The differences between one drop and another are very much perceivable to me, even as I myself am approaching that same sea to which others have returned, even as I spend my own return to that deep water considering the inevitable loss of my own individuality.

The raindrop analogy contains truth and comfort, but sometimes it feels like cheating.

It feels like that calm, poetic impulse of mine is saying, "Well, just imagine life and death as a more impersonal thing you don't care so much about—then it won't seem so grim."

But, of course, I do care about it. I want my life and my meaning to have weight and significance.

In light of my illness and history, I want to reject numbness and detachment even if it's dressed up as poetry.

And yet, death itself is as fundamental to nature, to what we are and how we arrived here, as sunlight or the grip of atomic bonds.

How do I reconcile this?

How do I honor the natural, earnest sadness and loss I feel in relation to death, while acknowledging that it's part of the same nature that gave me everything I treasure?

My depression asks me this question again and again and again.

I am a poet and my magic is in metaphor.

The raindrop metaphor is a potent sort of magic, but the tension comes when I try to employ it as an answer to the wrong questions.

This image of life as water returning to the sea is not an answer meant to address individual loss and grief, not directly. It's an answer to a broader accusation about nature itself being a cruel and purposeless taker. It's a rebuttal of the idea that our existence is somehow a crime scene and consciousness makes us all unfortunate witnesses. Nature isn't a crime or a thing of needless hardship. It is balance. It is creation and rest, blossoming and dormancy. Cycles and seasons, like water in all its many travels and native forms.

However, if I try to apply that same reasoning to soothe the loss of a friend, the metaphor suddenly feels awkward and disingenuous. That too is instructive. That feeling of tension says, "No, set aside grand thoughts of sea and sky. Zoom in. This loss is not about the broader nature of reality. It's about an absent voice, an empty seat at dinner, a person-shaped hole in your world."

In short, I have found that in my cosmology of personal meaning, death carries two kinds of sadness with two distinct calls to action.

One should be answered and refuted.

One should be held tight against my chest, accepted as part of the richness of life.

No, death doesn't mean nature is cruel and purposeless.

Yes, personal loss is real, and the pain that springs from it is rooted in our love, a love that shouldn't be ignored or dampened even though it costs us a river of tears.

Nature, its warm springs and biting winters, its rhythms of wakefulness and sleep, gave rise to our loves and our sorrows. I feel no contradiction in loving a life worth mourning. I feel no contradiction in being a raindrop that admires the sea and the sky, while embracing a wholesome grief for all of us who fall. My illness is defined by sadness, but not all sadness is an illness.

❖

Another vulture joined the first.

And another.

And another.

I looked on them and thought, *clergy*.

I thought, *wake* and *ritual*.

Vultures are such intelligent, gentle creatures and so much of what they are and do speaks to me of natural solemnity.

Turkey vultures do not kill.

Often, while searching the ground for a meal, they don't even flap their wings to stay aloft, trusting to columns of warm, rising air. Thermals carry the birds up and up in a widening spiral, while their vision and excellent sense of smell suss out carrion far below.

They are of the landscape and, somehow, a bit outside of it as well.

They stand at the threshold.

Shadows in the doorway.

❖

I feel a strong call to look kindly upon death.

Perhaps because it is such a key feature of all life and of the natural landscapes that ease my pain.

Perhaps because suicidal ideation has been such a prominent part of my mind and I feel obliged to be on good terms with all the regular inhabitants of my thoughts.

Perhaps because of the cyclical nature of my mental illness, my brain weather, building a kind outlook on endings is part of making the meaning I need to feel at home within my shifting self and my

uncertain world. My inner and outer ecosystems both feature conspicuous impermanence.

There are bones on my bookshelves alongside the poetry, novels,
comics, pebbles, pinecones, and dried flowers.

An elk vertebrae found and sent by a friend out west. A mourning
dove skull I spotted beneath the flowering arch of a bleeding-heart in
my garden. Other bits of the pale, hard scaffolding of animal life.

Why do I keep these things?

Why do they seem correct placed next to other beautiful tokens of
the natural world?

I keep them because they are indeed beautiful to me, because flowers
are lovely, but death also blooms.

These bones are earthen materials arranged and shaped by the passage of life the way flowing waters carve canyons or build stalactites.

Bones are natural, lovely objects.

Admiring a skull as a treasure of nature is emblematic of the way
I make friends with death without turning my back on its sadness or
meaning.

I can be sad about a personal loss, while also recognizing that death
itself is a cornerstone of the beauty of nature, part of the machinery that
drives everything I adore about life on this planet. Death doesn't need
to be either beauty or sadness. It can easily be both.

I believe that life, and this world, are worthy and wonderful without
adding any caveats about death.

Similar to my position that there is wisdom and comfort in considering the universe as an event more than a place, I find a parallel mindset
about the people I love and my own temporary self to be useful. We're
not just things, we are happenings.

Things may be lost.

Happenings are complete in and of themselves without needing to be possessed, unchanging, forever.

Do you judge a song substandard because it ends?

The ending is not the song, but the song requires it. Trying to continue the song forever would undo its purpose and character.

Yes, I'm here in my present, sitting at my desk speaking to you.

I'm also gone.

Scattered through the past and changed by the future.

It's happening even as I type these words.

It's even possible that you're reading this after my death.

Even while living, I was never one thing.

My identity was never a fixed point you could visit on a map. Not a place.

Neither is yours.

When we die, they may bury us or collect our ashes, but remember this: from baby teeth to skin cells, most of the matter that has worn your name is already spread throughout the world.

We bury our remains in the soil of our lifetimes.

This doesn't mean I always smile and wink at death. There's a reason I consider suicidal thoughts to be cause for medical intervention. Death can take care of itself. It doesn't need an ally in me.

My business here is life.

Yet, smile or not, death is there, and the earnestness I've learned to cherish tells me to neither ignore it nor curse its name.

Death is not a monster.

It is not a tyrant.

❖

Death, or perhaps lifelessness, seems to be the most common state of matter in this universe, the same universe that gave rise to ravens playing in a snowbank or wolves howling in the pines.

Is a thing diminished by its lack of life?

Somehow, that idea has never felt quite right to me.

When I lay my hand against a tree, I can certainly feel the life in it.

When I lay my hand against a stone, well, it does not feel dead to me.

I am suspicious of the idea that only matter and energy organized to be awake in the way we're awake is experiencing and participating in the universe in a vital and meaningful way.

I am similarly suspicious of the popular notion that we humans are how the universe "knows itself."

Don't misunderstand me, I adore inquiry and science, but the idea that human beings ushered new knowledge into existence when, for example, we named and measured a mountain feels a bit arrogant and simplistic to me, like arguing the word *oak* is a significant part of the trees it names. Is an oak enriched by our name for it? Is the mountain more itself once measured in meters? Does the universe really need us in order to know itself? That idea strikes me as presumptuous and self-aggrandizing.

We are not how the universe knows itself. We are how we know the universe. It's a small, but important distinction to make as we place judgement on what kinds of existence actually count, actually matter.

To be clear, it's not as if I think we're going to hurt the mountain's feelings. This is a distinction about us, about our meaning and understanding. This concept that human brains are the only, narrow path in existence through which valid knowledge may enter feels like a reductive and ungenerous understanding of the universe we call home. It also gives death more menace than it's owed. The end of a human life may halt one path to understanding, but not the only worthy path to understanding, to knowing.

Human ways of knowing are not the only ways of knowing.

We will never hear the shape of an oak the way a bat can through echolocation. We will never taste the presence of a vole beneath a woodpile the way a ratsnake can. We can't understand the crushing deep beneath the ocean as a physical sensation the way a sperm whale can. We are endlessly clever and sensitive creatures, but we do not have a monopoly on understanding. Our ways of knowing and perceiving are not the "correct" or "ultimate" ways, and embracing a bit of humility often is itself a path to deeper understanding.

Death is personal.

When I think about my death, I'm not thinking about all human understanding. I'm thinking about my understanding, my ways of knowing. I think that, while I'm not the only way the universe knows itself, I am a way and one day that way will be lost. I preemptively mourn the memories I hold dear. I mourn the rainbows cast by the light of reality passing through the prism of my unique perspective.

It is hard to imagine and hard to accept.

I do not control mortality, but I can shape my own perspective.

When my fear of death creeps in, I can have a conversation with that fear.

I fear the loss of the things I cherish about my life.

Yet, I know the things I love about myself do not exist solely within me. I don't need to trap them in stasis beyond time, walled-off and lifeless.

Empathy. Humor. Love. Curiosity. They will not be lost from the world when I return to the effortless unity of nature, when the raindrop of my life rejoins the sea.

They are of nature, the things I cherish within myself and within others.

The good in us awoke through natural processes.

Those processes will continue.

Those processes hold what we love in trust.

We guard life fervently because we love it and we fear the unknown of what is beyond it.

We guard it, but we do not own or control it.

Life is something borrowed.

A stolen moment.

A cool afternoon swim in a woodland lake.

That borrowed water of life is the same for humans and falcons and aphids.

The shape of the water matters to us, and that is sweet and natural.

But the shape is not what matters most.

It's the water. The lake. The fleeting afternoon.

Our sense of ownership can chafe against this talk of stolen moments and borrowed life. A grasping, little voice in us says, "I don't care that the good in life will continue. I care about what's mine. My life. My pleasures. My loves. I am the one being robbed by death."

It isn't so.

We may fool one another that individual ownership is a matter of law and finances, but nature is not fooled. Life is borrowed. It is magic held in common.

I suspect there are many kinds of consciousness, but there is only one kind of life.

The kind that's in you.

The kind that's in lichens and ferns and chestnuts and cities of coral dreaming in sun-rich seas.

One of the best qualities about the impermanence of everything is that our small moments of happiness can carry the same weight as any cosmic event of unthinkable time and distance. These things matter because we make them matter. Our choices matter.

And our thoughts? Our kindnesses? Our individuality?

Do they all vanish back into the gentle, unifying waters from which all life came or do they linger on in the waking world?

Well ... Nature does not waste, and the unknowable is as rich and sustaining as dark soil or autumn stargazing.

In lieu of having all the answers, I turn to what I do know.

In the end, you were a part of the sky on loan to a body.

A part of the sea that awoke to thought.

A part of the Earth who carried a name.

Even now, different from moment to moment as you are, the love and knowledge that you and your past selves set in motion lingers on in the world of your lifetime and beyond.

You are the mountain, but awake.

You are the rain, but breathing.

You are the forest, unanchored.

You are the soil, but with choice.

You are the sunlight, dreaming.

Soon, you will be these things again.

Earth. Rain. Forest.

So, what will you do until then?

❖

I departed before the vultures, retreating back into the trees as quietly as I could.

The idea of them stayed with me.

Dark birds standing in the periphery on the walk back to my car, on my drive back home, in the corners of the room as I tried to sleep.

That fall, I stood once again at a crossroads between hope and despair. The vultures that accompanied me there made no sound and said many, many things with the simple fact of their presence. Death was on my mind, but that didn't have to mean hopelessness and a pull toward desperate escape. It could mean a signpost pointing the way to natural transitions. It could mean standing in the presence of vultures, of holy people, gentle guides to doorways and crossroads, sunning their dark wings in an autumn clearing.

It was time to come to terms with the likelihood that my depression would not be solved once and for all. Not now. Perhaps not ever. It was also time to recognize that whether or not that was acceptable was entirely up to me. These were my paths to choose.

I hurt.

My pain, traditionally, drove me to consider death as a solution, an embrace of merciful oblivion.

But thoughts are not actions.

Feelings are not senses.

And now, more than ever before, I knew that my illness was indeed an illness and could be treated, influenced, and eased.

I wasn't broken.

I wasn't my brain weather.

And death, the teeth hidden within depression, hard as headstones—I didn't need to fear it.

I didn't need it at all.

I could consider it all I wanted, but I didn't need it. I had made peace with the idea that escape wasn't necessary.

I didn't need an escape because I had learned to accept myself, to be on my own side, to respect my own handcrafted meaning. I had learned to hurt without shame and without all-or-nothing condemnations of my life.

Once that was achieved, my mental landscape wasn't a place to flee. I wasn't in pain because I was a failure. My pain wasn't evidence that I was broken or beyond healing. Nor was I in pain because the world is pain.

Yes, I found hurt in the world, but I also found magic.

Wise herons waded the rivers, and clergy circled the skies.

I was mentally ill, but that was far, far from all I was or all the wide world had to show me.

Vultures are transitional creatures, so why shouldn't they stand with me at my autumn crossroads?

I nodded to those gorgeous, black birds, and I walked toward hope.

Their gentleness walked with me.

Chapter 20

HUMAN

THERE ONCE WAS A WEDNESDAY in January when I took a walk and saw a heron.

Like many of the most important moments in my life, I didn't understand its importance at the time.

I am still on that walk.

I believe, with effort and intention, I always will be.

Walking toward meaning, toward wonder, toward simple gratitude for nature and my kinship with it.

On a winter's afternoon nearer to today, I learned I was going to be a father.

The following fall, my son was born.

Arthur arrived in the small hours on a warm, overcast October night.

A day later, my dad sat in our hospital room holding his first grandchild, a tiny, tightly swaddled bundle in a white blanket.

Dad looked very ready and very happy to be a grandpa.

Stressed and sleep deprived, I asked him a question before I thought it through.

"It's been a very long time since you've been to a hospital for a happy reason, hasn't it?"

Dad had spent long hours in hospitals when my brother received his liver transplant.

He had spent long hours in hospitals during my mom's breast cancer treatments, and her subsequent decade of vicious, mysterious pain diagnosed as "new daily persistent headache," a condition which has never subsided.

More hours still as my mom faces early-onset dementia.

More for his knee replacement surgery.

My dad seems forever on duty, caring and advocating for his family, trying to stay ahead of illness and injury.

But not on that day.

That day, he was in a hospital to meet his grandson.

I saw the question I asked land, a gut punch, the sadness and the joy.

I saw him fight back the tears.

He just nodded.

Midwesterners being Midwesterners, I changed the subject.

We all learn to get by in this big, complex world as best we can.

Being human is a lot.

We are as natural as a gray squirrel or a flowering dogwood tree, but we are also different. We, like they, are unique. We make stories and then live inside them.

Some of the stories help us, some harm us.

Often enough, our stories become invisible and we forget that they are crafted things, that we can build and change them to serve our needs.

More than once, my stories, along with my illness, almost killed me.

But stories can change.

Here was a new story.

I was holding Leslie's hand as we sat smiling at our son in the arms of his grandfather.

A simple story to tell.

A story that would have been unimaginable a short time before, during the long, lightless span when my world was only pain and silent desperation.

And what saved me?

I did.

And Leslie.

And Dr. Bartman.

And Mark.

And the insights, work, and research of countless other minds.

And my parents.

And poetry.

And magic.

And little rivers and salamanders and cicada songs and the voice of crows and a raccoon prowling the shadows and deer walking through honeysuckle and the sight of a sugar maple standing green against a blue Ohio sky.

Storks don't deliver babies.

Herons might.

❖

I share my journey with depression and my love of nature with you because you and I are family and I want to add what strength I have to yours.

That's what humans do.

We build connection and community.

Sometimes bound together by place or blood or mutual need.

Sometimes bound by ideas and stories, by meaning and magic.

We share our strengths.

I am learning that *strength* is not as simple a word as I once thought. There are so many kinds of strength.

We need so many kinds, but not all are to be found growing naturally within ourselves. And so, we need each other.

For too long, I reached for strength defined as weathering pain and damage, weathering it on my own as privately as I could manage. But my long-term health and happiness were not served by this idea of strength. What I actually needed was gentleness. I needed connection. I needed community. I needed the bold vulnerability to seek help and the humility to honor other people's wisdom and expertise.

I needed to set aside ego and see value outside myself.

When I did, I found many hands outstretched to grasp mine, poised as if waiting for the moment I was ready to reach out.

Some from without.

Some from within.

Lifetimes of study and care dedicated to my illnesses and struggles reached back to me.

Unexpected friends and allies, perspectives and experiences beyond my own reached back as well.

Past versions of me, half forgotten, stood ready to remind me of what I loved before I feared loving the wrong things.

There are so many kinds of strength, so many kinds of knowledge and ways of knowing, and none of us possess them all.

I find teachers in surprising places.

Arthur is nearly four now and already he teaches me about appreciating nature.

Yes, I can write a poem about a bumblebee, but I've seen Arthur physically shake with excitement when he spots one. I do not have the skill to capture that feeling in words. I've seen him entranced and frozen with awe at the sight of a caterpillar or a sowbug. He can't quite distinguish left from right yet, but he tells me he is eager to visit the midnight zone of the ocean and see an angler fish. He speaks of such things with the reverence of prayer.

My son is a newcomer here and that, itself, is a kind of strength and knowledge, the potential energy of unwritten stories.

Humans are strange and wonderful animals.

❖

Kindness won't make you rich, but it will make you whole.

Kindness for each other, for ourselves, for our world.

I know there is hurt in your life.

I know there are tough times behind and tough times ahead.

These pains stick to us like burrs.

They tell us to lash out, to stop feeling, to turn away and turn inward.

But these impulses do not control us. They don't write our stories, and each time you hear them and answer, "No, not today," you have given a gift to the world.

Thank you.

Thank you for being here. Thank you for trying. Thank you for choosing kindness. Thank you for sharing yourself.

The world will give back to you in kind, but receiving those gifts can take a little practice.

Nature is out there and she is in you.

Meet her halfway.

As I write these words, the sun is shining and an image of a woodland path winding out of sight keeps tapping me on the shoulder.

It's a simple lesson that has taken me years to relearn.

When peace and wonder call, I go.

When the trees call, I go.

I believe something in the woods loves you.

One of those things is me.

ACKNOWLEDGMENTS

This book is for my son, Arthur. May your wonder stay beside you without thought or invitation. May nature never feel like a guest in your home or in your life. May curiosity, empathy, and kindness guide you in all things, especially in your relationship with yourself.

For Leslie. This book exists because of you, because you made space for me to heal and to grow, because of your selfless work and unshakable faith in your odd husband. You supported me when I could not support myself. You loved me when I felt unlovable. You reminded me that climbing trees and catching fireflies were not relics locked away in the past. You invited me back into the woods and back into myself. Thank you. I love you so.

For my parents, Molly and Dave Anderson. You make your world and your son better with your presence. You gave me a childhood full of nature and wonder and laughter, gifts beyond words of gratitude. If you need me, I'll be under the sassafras trees or up in my sugar maple. I will happily come when you call.

For everyone seeking a path to mental health. Your way forward may not be easy or tidy, but please trust that the path is there. You deserve to celebrate your efforts. I'm proud of you.

Thank you to my counselor, Mark Travis. You were right. I could write the damn book.

Thank you to my editor, Ryan Harrington, for your guidance, your enthusiasm, your critical eye, and your deadlines. I needed all of them.

Thank you to my agent, Rach Crawford, for your expertise, encouragement, and for humoring my insomnia while allowing me a glimpse of Australian sunshine during 2 a.m. video chats.

Thank you to my online community, my readers, listeners, and supporters. I live on a little street where I see more birds than people from my windows. Thank you for reminding me that I am connected to a global human family of kind, welcoming, generous people. I am sincerely grateful to share this amazing world with you.